Designing Intelligent Construction Projects

Designing Intelligent Construction Projects

Michael Frahm
Aalen, Germany

Carola Roll
Passau, Germany

Registered Office

John Wiley & Sons, Inc., 111 River Street, Hoboken, NJ 07030, USA

John Wiley & Sons Ltd, The Atrium, Southern Gate, Chichester, West Sussex, PO19 8SQ, UK

Editorial Office

9600 Garsington Road, Oxford, OX4 2DQ, UK

For details of our global editorial offices, customer services, and more information about Wiley products visit us at www.wiley.com.

Wiley also publishes its books in a variety of electronic formats and by print-on-demand. Some content that appears in standard print versions of this book may not be available in other formats.

Library of Congress Cataloging-in-Publication Data Applied for:

Paperback ISBN 9781119690825

Cover Design: Wiley
Cover Image: © Rachael Arnott/Shutterstock

Set in 9.5/12.5pt STIXTwoText by Straive, Chennai, India

Printed and bound by CPI Group (UK) Ltd, Croydon, CR0 4YY

C9781119690825_280722

Contents

Preface

Intelligence is the ability to adapt to change.

Stephen Hawking

Why a book with a focus on management cybernetics, lean management, and digitalisation?

Because cybernetics, as a mixture of the natural sciences and the humanities, teaches a holistic and universal understanding of the control and regulation of machines, living organisms, and social systems. Exactly the right thing for people who dare to think outside the box. Because lean management – which emerged in postwar Japan, a country that had to relearn industrial production in the face of huge unemployment, a lack of space, and being restricted by being an archipelago of many islands – brought a common-sense approach to industry, moving companies away from sluggish or unhealthy practices and conflict behaviour towards an attitude of trust and cooperation that focused on the product and the customer once more.

Because digitalisation is a great opportunity and the founders of cybernetics foresaw the technical possibilities of control and regulation regarding feedback and automation.

In addition to these changes, the combination of these topics offers incredible potential in terms of networking, enhancement, and feedback. The authors would like to mention here that the main purpose of this book is to show the interconnection between these disciplines. Methods and concepts are described concisely.

But what is an intelligent construction project organisation for us? In our understanding, construction project organisations are not only construction sites and construction companies but also engineering offices and client organisations, as well as other participants in the production of planning and construction processes that have an influence and must be considered. In our opinion, it is important to see the entire production system and its alignment and coupling internally and externally for a successful implementation.

For us, 'intelligent' stands for adaptability and robustness in order to produce products to the best possible extent, created according to logical and sensible processes and meeting the exact customer requirement. 'Intelligent' also stands for low-waste processes that promote the application of new technologies to reduce waste and put the cooperative collaboration in the foreground. Everyone knows (even if not everyone practises this) that the best way to work is cooperatively, that is with a bit of give and take. 'Intelligent' for us also means that we create production systems in which it is fun to work, in which there is a working culture of motivation. We are convinced that this is more intelligent than the 70-hour week.

In addition, 'intelligent' for us means the responsible and sustainable use of resources, as well as the use of technical and digital systems to relieve the human workforce of tasks that can be done better and more economically by nonhuman solutions.

This book does not claim to be exhaustive. Rather, the authors present topics that they consider relevant. This book is also not a research paper, but written for the interested user who wants to get to know and try out approaches in themselves and in their combination. Accordingly, we have intentionally avoided scientific terms where possible. Our aim is to make the text flow, to make the book an entertaining read, and so we have chosen its language, structure, and length with this very much in mind. Nevertheless, the book contains some theory that has to be relearned and internalised before it can be applied. And with complex questions there are usually no, or few, simple solutions. Therefore, the book avoids patent remedies. Practical examples, anecdotes, and questions for thought and reflection are offered instead. And we do not pretend to offer every answer, as we have also aimed to be succinct rather than exhaustive.

Both authors are German and so much of their environment and experience is German. However, whilst the book contains many examples from Germany there are also examples from around the world.

We have deliberately refrained from an epic approach to system theory and also the scientific discourse about definitions, terms, and which approach is ultimately the better one but have of course written about topics, areas, and approaches that interest us and with which we ourselves have gained experience. When this book talks about systems, it generally refers to organisations in an entrepreneurial context.

Donella Meadows, acclaimed author of *The Limits to Growth* (Meadows et al. 1972), and one of the most important 'systems thinkers' of her and our time (Meadows 2017), describes a system from elements, their relationships, and behaviour, and from its purpose or function. She explains in a generally understandable way that a system is, for example, a soccer team with the elements: players, coaches, the field, and the ball, whose relationships or connections are the rules of the game, strategies, communication of the players, and the laws of physics, whose purpose is to win games, play sports, or make money.

This understanding of systems also applies to other systems, such as a company, a city, an economy, an animal, a tree, a forest, which includes the subsystems trees and animals, the earth, the solar system, as well as the galaxy. Conglomerates without certain connections or functions are not systems.

Today, no one can avoid the design, analysis, and adaptation of systems, organisations, and processes. Understanding this is the key to dealing intelligently with increasing complexity.

Acknowledgements

We would like to thank Hamid Rahebi for making available his preliminary work, which he prepared as co-author for the German-language basis of this book and also for his support in the present expanded English version of this book. We thank DeepL, and especially Marc Beament who helped us with the translation into English.

Any remaining 'German English' we happily take the blame for!

We thank Wiley for their patience and support in the preparation of this book, Tim Bettsworth for copyediting it, and Paul Sayer, Amy Odum, and Mike New. We also thank Martin Jäntschke for his practical insights on implementing lean management in large organisational units working on large projects.

We thank all those from whom we were able to learn and continue to learn. Learning never stops.

About the Authors

Michael Frahm, born 1979. Educated in Stuttgart, Kaiserlautern, Saarbrücken in engineering and business law. Management courses at the HEC Paris, HHL Leipzig, and Northwestern University. Fifteen years of professional experience in mega construction project management. He is director of the nonprofit association for System and Complexity in Organisation (SCIO) for Germany and a Certified Advanced System Practitioner for this organisation.

Carola Roll, born in 1978. Educated in Straubing and Krems (Austria), in technical business administration, lean operations management, and integrated management systems. Over 20 years of professional experience in interface positions between business administration and technology in various medium-sized companies. She is director of the nonprofit association for System and Complexity in Organisation (SCIO) for Germany and head of Bavaria's related practice group.

1

Complexity, Cybernetics, and Dynamics

The Toyota style is not to create results by working hard. It is a system that says there is no limit to people's creativity. People don't go to Toyota to 'work' they go there to 'think'.

Taiichi Ohno

Designing Intelligent Construction Projects, First Edition. Michael Frahm and Carola Roll.
© 2022 John Wiley & Sons Ltd. Published 2022 by John Wiley & Sons Ltd.

In contrast to lean construction, cybernetics or a system-oriented approach is relatively unknown or unused in construction. Norbert Wiener, a mathematics professor at the Massachusetts Institute of Technology (MIT), coined the term 'cybernetics' in 1943. At the time, he was leading an interdisciplinary research project and was confronted with the problem of coordination and communication between different experts and disciplines. This was the official birth of a new science of communication and regulation. Hermann Schmidt, professor of control engineering in Berlin, is regarded as the founder of cybernetics in Germany. In addition to Wiener and Schmidt, other historical protagonists such as Heinz von Foerster, W. Ross Ashby, Humberto Maturana, Stafford Beer, Frederic Vester, and many others have lent significant meaning to the term 'cybernetics'. Meanwhile, cybernetics is used in different disciplines.

As described by Christoph Keese in 2016 in his bestseller *The Silicon Valley Challenge*, the understanding of cybernetics and system science is more relevant than ever. It serves as an essential model for digital transformation. Cybernetics is based on the idea that everything is connected to everything. It, therefore, encourages people to think out of the box – an essential characteristic in a networked world in which great importance is accorded to the effective confrontation of complexity and chaos.

Wiener, his colleagues, and successors would be delighted by the possibilities of today: digitalisation, the Internet of Things, and the chance to model and simulate systems with enormous computing power to make more and sounder predictions. Here is an introduction to cybernetics and systems science, which we think is essential to gain an understanding of the past, the present, and the future of (construction) organisations.

1.1 Complexity

The term 'complexity'[1] found its way into the language in the 1970s and has been defined in a number of ways since then. There are different approaches to and views on complexity. This reflects the subjective nature of complexity and that it depends on the context, actors, and observers. This section contains examples of 'complexity' from various fields.

1.1.1 Complexity in the Mathematical Sciences

In mathematics, 'complexity' is defined by the number of elements in the system and the variability of the feedback. The term 'complexity' is also associated with nonlinear system behaviour. Nonlinearity describes a system's sensitivity to even

---

1 Complexity: Latin *complexum*, participle perfect of *complecti*, 'to embrace' or 'to summarise'.

the slightest changes in the initial conditions. The so-called butterfly effect gives colloquial meaning to this behavioural phenomenon and explains that, theoretically, even the most minor changes in initial conditions (e.g. the wingbeat of a butterfly) can have a significant impact (hurricane) on the results. In geotechnics, construction mechanics, and structural analysis, the consideration of the nonlinearity of system behaviour is of great importance. In computer science, 'complexity' stands both for the computational effort required to solve a problem and for the information content of data.

Therefore, 'complexity' in the broader sense can be equated with calculability and system sensitivity.

1.1.2 Complexity in Sociology

In sociology, a distinction is made between factual, social, temporal, operative, and cognitive complexity. 'Objective complexity' describes the variety of different types of elements that can interact with each other. 'Social complexity' describes the interactions and feedbacks within the system. Extended by a temporal component, one speaks of 'temporal complexity'. 'Operational complexity' describes the fact that the system sets goals and that the system itself can bring about changes in the state. If there is a pronounced degree of controllability, we speak of 'operational complexity'. For Niklas Luhmann, the authoritative representative of sociological systems theory, complexity is an observer/observation-dependent fact and leads to a compulsion to select in systems (Luhmann 1994).

1.1.3 Complexity in Management

Hans Ulrich (Ulrich and Probst 1988; Ulrich 2001), a former professor of business administration at the University of St. Gallen, distinguishes between 'complicated' and 'complexity' as follows. He associates complicated more with the composition of a system, whereas complexity describes temporal variability more.

He expresses this as: 'Complexity is the ability of a system to assume a large number of different states in short periods of time. Machines are non-trivial systems whose behaviour is predetermined and predictable. Ecological and social systems are complex, "non-trivial" systems whose behaviour at certain points in time cannot be predicted' (Ulrich and Probst 1988).

As a rule of thumb, 'complexity' means that a system has many elements (E), relationships (R), and states (S) that change over time (t), as shown in Figure 1.1. An extension to this is 'chaos', which describes a state of complete disorder and independent causality. Much modern management literature has been based on Ulrich's understanding.

Following the acronym VUCA and Cynefin Framework, which has a solid link to complexity in management matters, were presented below for a better

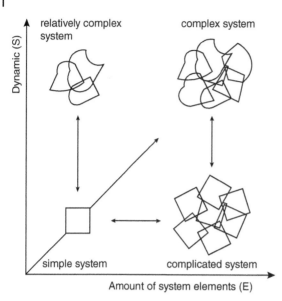

Figure 1.1   System states.

understanding. VUCA (Mack et al. 2016) or the VUCA world stands for:

- Volatility: e.g. frequent and rapid changes in the environment
- Uncertainty: not predictability of the future
- Complexity: many unknown elements exist internally and externally
- Ambiguity: information can be interpreted in different ways

It can be deduced from this that managers' previously tried and tested skills and abilities no longer endure in this new world and must be replaced by adapted leadership skills that are more strategically oriented and better suited to handling complexity (Lawrence 2013).

In response to the VUCA world, a VUCA acronym is again used. This is: vision, understanding, clarity, and agility.

The Cynefin framework (see Figure 1.2) provides an admired approach to reflecting complexity in a system context by Dave Snowden (Snowden and Boone 2007), a management consultant and researcher from Wales. *Cynefin* is the Welsh word for 'habitat' and is intended to reflect the point of view of the actor or observer on the context.

According to Snowden's Cynefin framework, a system will be classified between (Snowden and Boone 2007):

- simple
- complicated
- complex
- chaotic
- and disordered/confuse.

Figure 1.2 Cynefin framework.

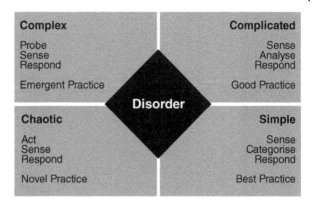

For each of these categories, a pattern of action is proposed. These are:

- **Simple system:** A simple system can be understood without further analysis and at the first attempt. Cause and effect are clear to all participants.
  The pattern of action: perceiving, categorising, and reacting is recommended. The existing facts are to be analysed, categorised further, and then implemented accordingly with a suitable procedure.
  Typical for this are tasks that can be implemented using predefined processes. This procedure is called 'best practice'.
- **Complicated system:** A complicated system is characterised by many cause-and-effect relationships. Cause and effect are no longer immediately comprehensible. A complicated system requires specific expertise and time to understand the elements in the system.
  The pattern of action: perceive, analyse, and react is recommended. This means that, analogous to the simple system, facts are to be explored, information is to be obtained, and expert knowledge is to be used on this basis.
  'Good practice' is recommended as the correct procedure. This means that there are various accurate solutions.
- **Complex system:** In a complex system, the cause-and-effect relationship can only be understood after detailed analysis and retrospectively.
  The pattern of action is: try, perceive, and react.
  'Emergent practice' is recommended. This means that a diverse approach is recommended, which considers a mixture of methods, working with cross-functional teams, and experimentation.
- **Chaotic system:** In a chaotic system, it is not predictable how small changes in the initial conditions will affect the system's behaviour in the long run.
  The pattern of action act, perceive, and react is recommended.
  For the chaotic system, Snowden advises the use of just a handful of authorised people to act to achieve an immediate effect and stabilise the system and

Table 1.1 Cynefin framework (PDCA).

	PDCA circle			
	Plan	**Do**	**Check**	**Act**
Simple systems	Sense	Categorise	—	Respond
Complicated systems	Sense	Analyse	—	Respond
Complex systems	—	Probe	Sense	Respond
Chaotic systems	—	Act	Sense	Respond

manoeuvre it into another system state. He calls this 'novel practice' (Snowden and Boone 2007).

- **Disordered/Confused system:** The system cannot be classified by the actor or observer in confusion. In this case, the task can be broken down into smaller tasks, for example.

  In the case of disorientation, the system cannot be classified by the assessor. In such situations, people often withdraw to their comfort zone without assessment and make decisions based solely on their own experience. This is not necessarily wrong. Referring to 'intuition' and the researcher Gerd Gigerenzer's (2008) research results, we see that intuition can also often lead to good and quick decisions in complex and chaotic situations.

In general, the Cynefin framework supports system understanding and classification and provides a recommendation for acting.

Mapping of the Cynefin framework with the Deming circle[2] (Plan, Do, Check, and Act) is shown in Table 1.1.

1.1.4 Complexity in Construction Management

Patzak (2009) and IPMA (2016) provide a complementary approach for the classification and partly also for measuring complexity in project management. The procedure proposed by Patzak is cumulative and uses a scoring table and includes the following areas to be scored:

- project goal
- project subject

---

2 The Deming circle, or PDCA circle, describes an iterative four-phase process for learning and improvement devised by Walter Andrew Shewhart, a US physicist. The origins of the process lie in quality assurance (Deming 1986).

- project task
- project executors
- project environment.

The complexity of the building industry is determined by its institutions and actors. Many participants at the administrative and operational level make for a great deal of complexity. This is intensified by a high cost and time pressure. Dirnberger (2008) notes that complexity arises at the interfaces since even small construction projects today involve more than 100 participants.

Schwerdtner (2007) distinguishes between organisational and technical complexity and takes into account the unique nature of construction projects. 'Technical complexity' refers to the building structures, and 'organisational complexity' relates to the systems they create and operate.

Hoffmann 2017 distinguishes between:

- the complexity of the building (object)
- the complexity of the processes (project)
- the complexity of the relationship structure (behaviour).

In the case of the long term (10 years or more), changes in standards and the legal situation are also relevant. Also, the growing critical interest for a project increases its complexity. Just-in-time planning, lifecycle orientation, the decreasing level of trained site personnel, conflicting implementation cultures, competitive constraints, and the fact that construction is often started on a greenfield site are all factors that increase a project's complexity.

For the purposes of classification, specifically for construction project management, complementary approaches were developed by Lechner (2015) and especially by Hoffmann (2017).

1.1.5 How to Cope with Complexity

In the context of complexity, understanding what it is and what approaches can be implemented is essential. Specific approaches are suitable as already described, e.g. with the Cynefin model.

According to Dietrich Dörner (2012), the following are common mistakes in dealing with complexity.

- First error: Wrong target description. Individual objectives are worked on and the overall system and its effects are disregarded.
- Second error: No networked analysis. No order principles are created to evaluate data or large amounts of data, e.g. using feedback loops.
- Third error: Wrong emphasis. One concentrates, e.g. only on one pathological focus; other problems are disregarded.

- Fourth error: Side effects are ignored. One works single-mindedly on a problem without caring about the side effects.
- Fifth error: Tendency to override. If small interventions show no effect in cases of maladministration, a system is heavily intervened in. If there is a time delay in accumulating unexpected results, the override is completely stopped.
- Sixth error: The tendency to authoritarian behaviour. Whoever has the power in the system thinks they have seen through it. This can be fatal if you think you are in control. It is more important to activate self-regulation in the system.

Jurgen Appelo (2010), developer of the Management 3.0 approach, recommends the following:

- Address complexity with complexity (see also Ashby's Variety in Section 1.2.2).
- Use a diversity of models.
- Assume dependence on context.
- Assume subjectivity and coevolution.
- Anticipate, adapt, and explore.
- Develop models in collaboration.
- Copy and change.

The following heuristics are also the first guidelines in dealing with complexity:

- Take the best: Using exclusion strategy – exclude irrelevant information.
- Tit for tat/tit for two tat: Using cooperation strategy – reduction of complexity (disturbances) through conflicts (see also Chapter 4, Section 4.7.2).
- Simple frame: Through a framework strategy (e.g. 10 commandments), means face/select/ignore complexity through the definition of a simple framework.
- Pareto strategy: Using the Pareto principle (80/20) to counter complexity through effort and benefit.

For construction management, the kpbm° heuristic of Frahm (2015) can be taken into account. It concentrates on the following system characteristics to build a robust system behaviour for construction endeavours.

- **Viability:** Viability is to be understood in the entrepreneurial sense. This means:
  - o Adaptability to change.
  - o Ability to influence and shape the environment.
  - o Ability to open up new environments.
  - o Ability to make a positive contribution to big picture delivery.
    Viability expresses that construction projects should be designed so that there is sufficient capacity to function effectively as an organisation internally and with its environment externally.

- **Attenuation and amplifying:** Interrelationships come to the fore by attenuation and amplifying. See, for example, Section 1.2.2.
- **Bottleneck concentration and flexibility:** Key elements in the planning and construction process are resources. Bottlenecks in this respect must be identified and overcome. Furthermore, appropriate buffers have to be considered.
- **Cooperation:** Cooperation is the superior strategy for achieving something together and achieving the best for everyone involved.

After all these explanations, it is clear that dealing with complexity is not an easy task. Umberto Eco (1988) writes in his novel *Foucault's Pendulum* something of a metaphor: 'For every complex problem, there is a simple solution, and it is wrong.'

Often there are no simple rules to cope with complexity. As a rule of thumb, one can say, the more complex the situation is, the greater the mix of people, disciplines, and methods is needed.

1.1.6   Interaction and Autopoiesis

Interactions are essential elements in systems. The Chilean biologists and neuroscientists Humberto Maturana and Francisco Varela (1992) present the concept of structural coupling to a broad public in their book *Tree of Knowledge*. Structural coupling occurs in interactions, which means the structure changes the environment evolutionarily, and vice versa.

There are many possibilities for design, and the more accurate a project becomes, the more supporting and hindering forces are released. Within the framework of this design process, a structural coupling must take place. The management of a project requires a stable project environment. A vital foundation stone is laid in the design process.

The management of construction projects happens mainly in the beginning and with large-scale projects is often more a social process than the implementation of a technical project. Those responsible and owners quickly realise that technological and economic professionalism alone is not enough. The formation of coalitions, political connections, and public perception are essential areas in the process.

The task of forming projects before they can be planned, built, and operated on can be an extraordinarily disorganised and complex process. This is expressed by the process of coordination and design that takes many years or decades with the slow and challenging consensus building. Through many small steps, the integration of the system into the environment takes place. Many independent questions cannot be solved simultaneously. Issues that have already been solved and defined are often restarted through long design processes, for example when marginal conditions and political power relations change, even if these have already been formally decided.

This evolutionary mechanism reflects the Darwinian principle from 1859 of the interaction between variation (specialisation) and selection (adequacy of the most suitable), which is still valid today (Darwin 1859). The relationship is coupled when recursive interactions have reached stability, and further changes continue for the long run co-evolutionarily and in the same direction.

With the look at structural coupling, you can answer operational and strategic questions for your organisation and your project alike, what a specific relation does, where it takes you, what identity can be created out of this, and how can you build stability?

Maturana and Varela also established the concept of *autopoiesis* (ancient Greek for 'self-creation'). Autopoiesis describes systems that refer to themselves, and create and maintain themselves out of themselves. The basis of their self-organisation is always directed towards a state of equilibrium and thus towards self-preservation. When a form of stability is reached, the system is structurally coupled. This can be repeated many times.

An example of an autopoietic system approach is the viable system model (VSM).

1.2   Viable System Model

The Conant–Ashby theorem (Conant and Ashby 1970; Conway 2021), also known as the Good Regulator, means:

> Every good regulator of a system must be a model of that system.

This means that the management of an organisation can only be as good as the model on which it is based. Conway's law gives the Conant-Ashby theorem practical meaning. Melvin Conway, an American computer scientist, was made public in 1986 and became more widely known in digital transformation and the implementation of software projects. It reads:

> Organizations which design systems … are constrained to produce designs which are copies of the communication structures of these organizations.

This means that, in the case of insufficient communication between departments, defects in the product manifest themselves exactly where the interfaces do not work. Products are, therefore, results of the communication structures of their organisations and thus the results of the underlying models. A Harvard study confirmed Conway's law (MacCormack et al. 2008). To map the behaviour of systems in an organisational context, a suitable model is recommended.

As described already, construction projects are usually complex projects. To counter 'complexity' and 'chaos' in the entrepreneurial context and create stability, management cybernetics was developed by Stafford Beer (1979, 1995), a professor of business administration and operations research at the Business School of Manchester. He transferred cybernetics approaches to companies and the business world generally.

Beer's most well-known application is the VSM. As an alternative to a hierarchical organisation structure, Beer oriented the development of the model on the successful model of evolution in terms of viability: 'the central nervous system of mammals'. The VSM attempts to balance and order the system and provides a fractal (see Figure 1.3) or self-similar structure. This means that, on the one hand, a balance between control and autonomy in an organisation is striven for. On the other hand, a fractal structure at all organisational levels exists.

The same or similar generic organisational code can deal with complexity and chaos in a relatively simple way and help create order. The VSM serves as a basic structure or as a map with which one can orientate oneself.

The rule is:

The purpose of the system is what it does – keyword **POSIWID**

This means that production is the reason the organisation exists. The organisation must follow the production process. As the organisation can exist out of more than one viable system, there can be more than one purpose.

1.2.1 The Static Perspective on the VSM

The static perspective of the VSM is about analysing existing systems with the help of it.

Thus, in addition to the viability of a system, any existing problems in the corporate structure and the structural and procedural organisation can be examined.

The goal of this is to identify cybernetic approaches and missing elements about cybernetics (as with a checklist), if present, and use them as a basis for the conceptual design of an optimised system.

This has the advantage that there is no need to worry about forgetting anything.

The idea of making the VSM equally useful as a diagnostic tool goes back to its creator, Stafford Beer, who in his 1985 book *Diagnosing the System for Organizations* provides the reader not only with a description but also with a workbook in the best sense with concrete working instructions. At the same time, Beer is aware that this approach and its presentation are entirely new (Beer 1985). In his chapter 'A Cybernetic Method to Study Organizations', Raul Espejo explicitly emphasises

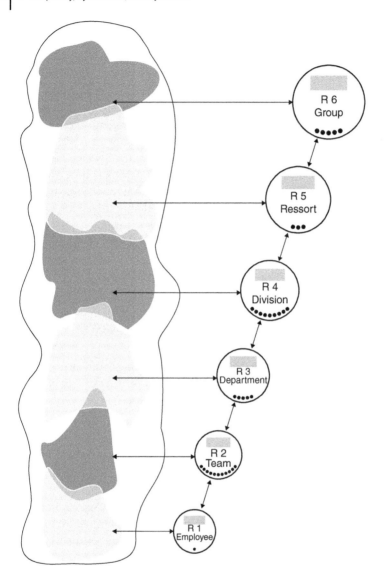

Figure 1.3 The fractal structure.

that the VSM, although perceived mainly as a diagnostic tool, also opens up other possibilities (Espejo 1989).

The VSM consists of six horizontal system levels (see Figure 1.4):

System 1   Operation
System 2   Coordination

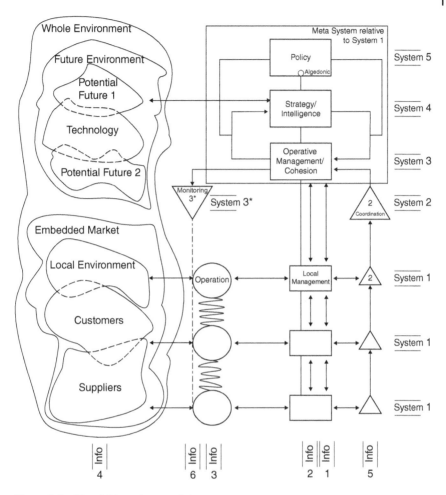

Figure 1.4 The viable system model.

System 3   Control and Cohesion/Operational Management (inside and now)
System 3*   Audit and Monitoring
System 4   Intelligence/Strategic Management (outside and then)
System 5   Policy/Normative Management

and six vertical information channels:

Channel 1   Intervention and Regulate
Channel 2   Allocation of Resources
Channel 3   Operational Interrelationships
Channel 4   Interrelationships of the Environment

Channel 5   Coordination (Sympathicus) = System 2
Channel 6   Monitoring (Parasympathicus) = System 3*

Furthermore, the VSM includes an algedonic channel and transducers. Algedonic signals are alarm signals that transmit either positive or negative messages directly into system 5. Transducers are converters that form the interface between the subsystems. They ensure the maintenance of information authenticity.

1.2.1.1   System 1: Operation

System 1 (S1) consists of three elements:

- environment
- operation
- management.

These three units build the:
'System in focus', in other words the reason the organisation exists.

In this system, all the main activities of an organisation are collected. All three elements interact with each other. The entire company must always pay great attention to this system level.

Examples of an S1 are producing units such as construction teams, construction companies, or designers.

It must be noted here that organisations may consist of several S1s. These usually represent strategic business units, product lines, or similar. When modelling a VSM, urgent attention must be paid to the viability of the S1s.

1.2.1.2   System 2: Coordination

System 2 (S2) has two tasks. First of all, it represents a communication medium between S1 and S3 through a standardised processes, a common language, and it coordinates all S1 systems with each other. Secondly, it is the institutional place where self-organisation takes place. It has an activating, or sympathetic, effect.

Examples are daily short informal meetings, operation and production plans of any kind, scrum/Kanban boards, project control, production planning and control, six sigma, the last planner system, or takt planning and control.

1.2.1.3   System 3: Operational Management

System 3 (S3) deals with the present business of systems 1–3 and must make all operational activities as efficient and effective as possible. It allocates resources and demands results. S3 competes with S4 for resources and receives normative specifications from S5.

Examples for an S3 instance are project, team, or department leader, or a managing director.

1.2.1.4 System 3*: Monitoring/Audit

System 3* (S3*) is a review channel with an institutional absorbing, or parasympathetic, effect.

An example of this subsystem is a regular walk by management around the construction site, having conversations with the foreman and workers to get an additional picture of the production process or a project audit with an objective view from the outside into the project process. Lean tools such as 5S audits or generalist layered process audits can also be located here.

1.2.1.5 System 4: Strategic Management

System 4 (S4) focuses on the strategic issues of the overall organisation and deals with future problems from the environment. It competes with S3 for resources.

Examples of S4 are strategic purchasing or developing new business areas such as a building information modelling (BIM) strategy for the entire company.

1.2.1.6 System 5: Policy

System 5 (S5) represents the identity of the organisation. Topics such as values, norms, ethics, and culture are addressed here and transferred into the organisation. It is the highest decision-making unit, makes fundamental choices, and, if necessary, regulates between S3 and S4.

Examples are corporate values and principles. These organisational principles are reflected in all subsystems. The highest decision-making unit here is the board of directors. The vertical information channels are, according to Steiner and Jernej (2017):

- **Channel 1 – Intervention and Regulate**
  These central project specifications must be known from S5 to S1 (e.g. project manual, construction contracts, engineering contracts, building permits).
  Question: What is expected, and what are the rules?
- **Channel 2 – Allocation of Resources**
  This involves allocating resources (e.g. personnel planning, resource planning, contractual regulations).
  Question: What is needed? In what way are the resources allocated? And how is this allocation justified?
- **Channel 3 – Operational Interrelationships**
  This is the channel of inner relations, which often informally occur but can be essential for positive functioning (e.g. celebrations, rituals, or private acquaintances). In other words, it is the culture of how to build.
  Question: What culture exists between the units?

- **Channel 4 – Interrelationships of external and internal environments**
  Management and producing units are interlinked with different environments.
  These environments are connected in different ways and influence each other.
  They are relevant for the present and future (e.g. stakeholders, public, politics).
  Question: How do environments affect each other, and what are the consequences for the system?
- **Channel 5 – Coordination (Sympathicus)**
  Communication of all S2 systems.
  Question: How do coordination and control take place?
- **Channel 6 – Monitoring/Audit (Parasympathicus)**
  Communication of all S3* systems.
  Question: What happens in production?
  Are the results reported to S3 via channel 6 congruent with the communicated results from S1 via channel 2?

The authors are aware that the preceding observations and explanations can only provide a brief, albeit concise, introduction when measured against the depth and complexity of the VSM. If you are interested in further and very detailed considerations of this topic, the authors refer to the three-volume *The Viability of Organizations* by Wolfgang Lassl (2019/2020).

1.2.2 Ashby's Variety

The VSM is the theoretical and practical application of Ashby's law. Therefore, an essential basis for developing a deeper understanding of the VSM is the knowledge and experience of Ashby's law, respectively Ashby's variety theorem or Ashby's law of requisite variety. Ashby's variety theorem makes a fundamental contribution to this, it says:

> Only variety can destroy variety.

Or:

> Only variety can absorb variety.

Figure 1.5 shows this theorem transferred to the management paradigm.

The variety of the environment (V_e) far exceeds that of the management (V_m) (see Eq. (1.1)). For construction projects, it can be concluded that the environmental variety (V_e) is generally more significant than the behavioural repertoire of construction management.

$$V_e >>> V_m \tag{1.1}$$

Figure 1.5 Varieties 1.

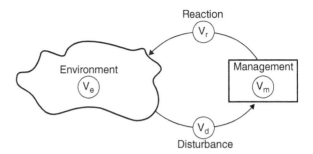

Ashby (1985) uses the variety as a measure of complexity and uses his variety theorem, which represents the variety of consequences (V_c), as the quotient of the variety of disruption (V_d) and the variety of system control (V_r) (see Figure 1.6). If this equation is in equilibrium, then there is stability (see Eq. (1.2)). If the variety of consequences (V_c) is larger or much larger then we have ultra-stability and the system can compensate changes without crisis (see Eq. (1.3)).

$$V_c \geq \frac{V_d}{V_r} \text{ stable} \tag{1.2}$$

$$V_c > \frac{V_d}{V_r} \text{ ultra stable} \tag{1.3}$$

A system collapses when the disturbances are significantly more meaningful than the system reactions. The variety theorem then becomes unstable (see Eq. (1.4)).

$$V_c < \frac{V_d}{V_r} \text{ unstable} \tag{1.4}$$

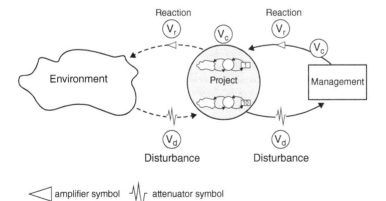

Figure 1.6 Varieties 2.

According to Markus Schwanninger (2006), former management professor at the University of St. Gallen, the variety can be calculated as described in Eqs. (1.5) and (1.6).

$$V = m \times \frac{n \times (n - 1)}{2} \tag{1.5}$$

Element-oriented equation:

$$V = z^n \tag{1.6}$$

where V = variety, m = number of relationships between each pair of elements, n = number of elements, and z = number of potential states for each element.

It doesn't matter which equation is chosen; variety always follows the law of power. It grows exponentially with the number of elements and the number of relationships, as shown with Eqs. (1–5) or with the potential of states, as shown Eq. (1.6).

In 2011, Frahm introduced two measures:

- variety number
- degree of variety.

Both quantities serve to measure complexity in organisational systems and are defined here.

1.2.2.1 The Variety Number
This describes the quotient of the sum of all interrelations (relationships) W of a project structure to the number of order levels OE (see Eq. (1.7)).

$$VZ = \frac{\sum W_{ij}}{\sum OE_{ij}} \tag{1.7}$$

1.2.2.2 The Degree of Variety
This is the quotient of the sum of all interrelations W to the number of nodes (elements) N of the systems (see Eq. (1.8)).

$$VZ = \frac{\sum W_{ij}}{\sum N_{ij}} \tag{1.8}$$

Udo E.W. Küppers (2019) summarises this as:

> If we consider, for example, our living nature with tens of billions and more interactions between organisms on the one hand and organisms with inanimate nature on the other, as well as levels of an order extending from atomic to biospheric space, the sheer number of interactions alone provides an unimaginably high number of varieties.

We first have to learn and understand how to handle the complexity of systems – of whatever kind – correctly. The core of this learning and understanding is networked, systemic thinking, and acting. Without this, any – even more sustainable – solution approach in the environment of complex structures and procedures is doomed to failure.

Social structures remain controllable if the steering system has the same variety (in other words the exact scope of actions) as the system itself. This approach gives rise to two possibilities:

- On the one hand, the internal variety of a system.
- On the other hand, the external variety of the environment.

Internal variety can be amplified to increase the scope of actions. Attenuation can reduce the external variety of the environment, which affects the system.

Example Owing to the rapid progress of a project, a senior site manager with a team of 20 members complains about excessive information from meetings, emails, letters, reports, telephone calls, etc., with customers, construction companies, authorities, and subcontractors. Ultimately, they no longer knows where his head is at. A classic case of information overload. Or, as Ashby perhaps would say: 'The external variety of the system is greater than the internal variety of the senior site manager.'

The senior site manager now has two options: they can 'amplify' or they can 'attenuate'.

We have some examples of systemic amplifiers and attenuators. It should be noted that sometimes attenuators and amplifiers can do both functions depending on the system perspective (see Tables 1.2 and 1.3).

Now imagine the following theoretical situation: you observe two contracting parties during abstract contract negotiations. One party should be called R (regulation) and D (disturbance). The negotiation strategy is based on the following table. If R succeeds in achieving an 'a' in the negotiation based on the table, they have won the negotiation. Table 1.4 is known to both contracting parties before the negotiation.

D begins the negotiation and must select a number, then R follows and selects a Greek letter. The intersection of both moves is the result. If R can get an 'a' with it, they win.

It quickly becomes apparent that R always wins with this theoretical negotiation configuration. R can consistently achieve an 'a' no matter which number D plays. R is superior to D in this case; they can fully determine any result they need.

What happens now if you go into supplementary negotiations with a different configuration? Let's assume that the same rules apply as before, but the negotiation is based on Table 1.5.

Table 1.2 Amplifier.

Amplifier:	
Structural measures	**Attention**
Integrated teamwork	Unclear responsibilities
Built-up resources	Incorporation losses
Outsourcing	Integration problems
Capacitive measures	**Attention**
Enlarge staff department	Water head
Best practise, experts, consultants	Loss of competitive advantage
Informative measures	**Attention**
Conferences	Ineffective conferences
Training of employees	Ineffective training
Real-time information	Flood of information

D has five options in this negotiation configuration, and R has four. If 'a' is the goal for R again, then R can always win the negotiation, just like in the example above; they can consistently enforce this result.

But if 'b' would be the negotiation goal for R, then R cannot always win. If D, for example, would choose line 2 or 3, then R could not force the result because 'b' does not occur in its respective configuration. In this case, R would lose to D.

The quintessence of this theoretical example is that each system configuration can create various states and possibilities. The tables can be understood as a simplified representation of the environment or internal and external conditions of a construction project. R can only steer D if it has a wider variety.

Based on that, Ashby derived his law of requisite variety, which says:

> Only the variety in R can reduce the variety in D; or in other words: only variety absorbs variety.

This law is universally applicable.

It also teaches us that the control capacity of R cannot be greater than the capacity of its transmission channel. Here is an analogy to the Shannon[3] theorem from communications engineering.

The following example is intended to illustrate this situation. It is supposed to highlight the regulation and regulation capacity.

---

3 Claude Elwood Shannon (1916–2001), American mathematician and electrical engineer. He is regarded as the founder of information theory.

Table 1.3 Attenuator.

Attenuator	
Structural measures	**Attention**
Division by structure or product	
Specialisation by markets	
Function-oriented structure	
Delegation	
Micromanagement	
Capacitive measures	**Attention**
Short-term planning	Long term is neglected
Long-term planning	Short term is neglected
Bottleneck concentration	Concentration on wrong bottleneck
Management by objectives	Adaptability can be lost
Informative measures	**Attention**
Strong administration	Loss of creativity
Management auditing	Loss of efficiency
	Feeling of control

A project manager is responsible for a construction project of six subprojects. Each subproject grows every day with an internal variety of 10^6[4] bit. The project manager receives information from his six subproject managers (SPMs) and a three-person project controlling team (PS). Each group can fictitiously report

Table 1.4 Negotiation 1.

		R		
		α	β	γ
	1	c	a	b
D	2	a	b	c
	3	b	c	a

---

4 The term bit (binary digit) is used in computer science, information technology, communications engineering, and related fields as a unit of measurement for the information content (see also Shannon, Nit, Ban). One bit is the information content which is contained in a selection of two equally probable possibilities.

Table 1.5 Negotiation 2.

			R		
		α	β	γ	δ
D	1	a	a	b	d
	2	d	a	d	a
	3	a	a	a	d
	4	b	a	b	d
	5	d	b	a	b

60 letters per minute to the project manager for eight hours every day. The transfer rate of reporting is two bits per letter (L).

The question arises: Does the information channel have sufficient capacity to enable the project manager to steer the project?

The variety of development that affects the system is (see Eq. (1.9)):

$$V_d = 6\,000\,000 \text{ bit} \tag{1.9}$$

The capacity of the information channel is (see Eqs. (1.10) and (1.11)):

$$V_r = (6 \text{ SPM} + 3 \text{ PS}) \times \left(60 \text{ L} \times 60 \text{ min} \times 8 \text{ h} \times 2\frac{\text{bit}}{\text{L}}\right) \tag{1.10}$$

$$V_r = 518\,400 \text{ bit} \tag{1.11}$$

The variety of information that generates the project's development is about 11.5 times that of the information provided by the SPM and PS. The capacity of the information channel is not nearly sufficient.

If you want to be well prepared to deal with such complexity, you must process the information adequately.

What does the situation look like if the project manager can give orders at 400 bits per minute and 10 hours per day? Let us assume they have complete knowledge of the construction project. Would the capacity of the command axis be enough for complete control? For this purpose, we consider the following calculation (see Eqs. (1.11) and (1.12)):

$$V_r = 400 \frac{\text{bit}}{\text{min}} \times 60 \text{ min} \times 10 \text{ h} \tag{1.12}$$

$$V_r = 240\,000 \text{ bit} \tag{1.13}$$

The capacity is equivalent to approximately 1/25th of the variety resulting from the development of the construction project. This theoretical explanation shows the gap between the variety of disturbance and system control.

But there are also simpler examples than these more theoretical ones. This would be, for example, the game of chess. At the beginning of the game both players have the same pieces (= variety). This changes when players lose a piece (their castle or queen, for example). Or if we take football, when both teams have 11 players on the field, the variety is balanced.

Ashby has not been the only one to deal with this topic. An alternative is the selectivity strategy developed by Niklas Luhmann(1994).[5] It says that a system can only be controlled by selection. It follows from this that complexity forces the compulsion of choice. According to Luhmann, it is unnecessary to develop control systems on the same complexity level as the environment. In contrast, to decrease the complexity of the environment, control systems need a simple structure. Selection strategies are necessary to choose between potential system modes.

It seems that the approaches of Luhmann and Ashby are contradictory, but we would contend that Ashby was often misunderstood. It should not be the case that management should have so much self-variety that it is no longer practicable. It is more important to choose a manageable environment. It is essential for a manager to gain enough behavioural repertoire to cope with the existing complexity of their working environment.

1.2.3 The Dynamic Perspective on the VSM

The views in this book are mainly based on and influenced by Patrick Hoverstadt (2009, 2018, 2019), Global Chair of the SCiO – System and Complexity in Organisation and Managing Director of Fractal Consulting.

The VSM is not just a static or a mechanical model, as often seen or perceived. When people first come across it, it looks a lot like a schematic of a radio or a television set. But on closer inspection it will be seen more as a dynamic approach to understand and follow the variety of flows and their changes over time in the system.

As explained elsewhere in this book (see especially Section 1.2.2), systems and, in this case, organisations and projects must be in a stable condition with themselves and their environment. This will be successful if you meet the requirements of Ashby's law, operate in the right environment, and have sufficient capacity (required variety) to counter the variety of change with the appropriate variety of system control in the proper horizontal system levels and vertical information channels.

The variety flows on the horizontal axis given in Figure 1.7 provide autonomy, and the variety flows with the vertical axis provide the entire system's cohesion. According to Martin Pfiffner (2020), Chairman of the Foundation Oroborus and author of *Die Dritte Dimension des Organisierens* (*The Third Dimension of Organising*), senior management controls the systems 1 on the vertical axis, where they

---

5 Niklas Luhmann was a German sociologist. As a German-speaking representative of sociological systems theory, Luhmann was one of the giants in the field of the social sciences in the twentieth century.

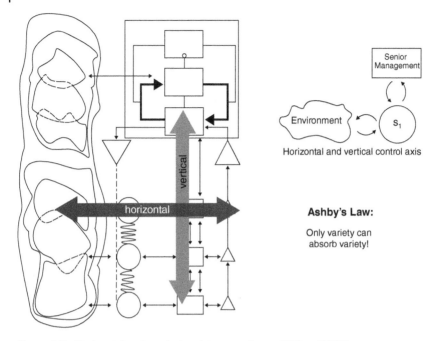

Figure 1.7 Horizontal and vertical variety according to Pfiffner (2020).

are not able to control themselves. Systems 1 should strive for autonomy; the more they do so, the stronger the variety of the vertical axis or senior management must be. To satisfy Ashby's law, the horizontal and vertical axes must be in balance.

The variety of the horizontal axis is determined by the number of systems 1, their diversity, and their ability to control themselves. The variety of the vertical axis over the six information channels is described in Section 1.2.1. The six vertical channels together must be able to handle the horizontal variety.

If we speak of variety flows, we speak qualitatively about the ability to handle the quotient of the variety of disturbance and system reaction present in every single relationship and the capability of the organisation to cope with the given variety and the given complexity.

With the dynamic perspective of the VSM, the variety in systems must be read more like a weather map with low- and high-pressure areas, which change over time. This means you must have to look out for:

- flows
- variety balances
- amplifiers and attenuators
- order and disorder in organisational matters.

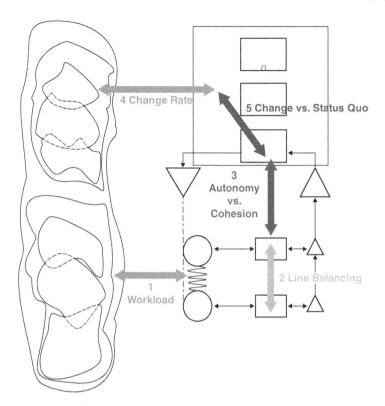

Figure 1.8   5 Variety balances.

For a deeper understanding of the VSM's dynamic perspective, five significant varieties of balance are presented. These are (see Figure 1.8):

- workload
- line balancing
- autonomy vs. cohesion
- change rate
- change vs. status quo.

These balances appear simultaneously with the subsystems.

The subsystems of the VSM often evolve in the following order with the following tasks:

- System 1 = The production, the job to be done, the purpose.
- System 2 = We need some coordination, standards, and reporting.
- System 3 = We need to be efficient.
- System 4 = What happens in the future?
- System 5 = What is our identity, vision? And why?

1.2.3.1 Variety Balance 1: Workload

With variety balance 1, the critical relationship between the production (systems 1) and its operational environment (i.e. market, customer) is shown. On this production system level, it can be asked: What are our complexity drivers, and how can we manage the variety created through the relationship with the environment?

Questions include:

- Can system 1 manage the existing variety (workload) from its environment?
- Can it manage its existence in the long run?
- Is system 1, according to the principle of subsidiarity,[6] able to handle as many tasks as possible directly?
- Can it adapt to a changing environment?
- Is it financially affordable and desirable that system 1 acts relatively autonomously?

1.2.3.2 Variety Balance 2: Line Balancing

This variety balance is concerned with the coordination of production systems with system 2. This information channel is one of the strongest to adapt to vertical complexity. According to Pfiffner (2020), a well-adjusted system 2 performs its task 70% via self-coordination, 20% via active coordination, and 10% via the individual decisions of system 3. The importance of this variety balance is often chronically underdeveloped.

Questions include:

- Is there enough standardisation, i.e. between the program and the project level?
- Is there an adequate resource allocation between several system 1s in the context of variety balance 1?
- Is there enough maturity in the processes and production planning?

When there is silo thinking between the departments, it appears here and leads to, for example, turf wars and a bunker mentality inside the organisation.

1.2.3.3 Variety Balance 3: Autonomy vs. Cohesion

With variety balance 3, two balances compete.

The first is resource versus performance. First, you see whether the loop between resource supply and results is closed.

---

6 The principle of subsidiarity means that higher, i.e. state, institutions should only (but constantly) intervene relatively if the possibilities of the individual, a smaller group, or lower hierarchical level alone are not sufficient to solve a particular task. In other words, this means that the level of regulatory competence should always be 'as low as possible and as high as necessary'.

Questions include:

- Are there enough resources in system 1 for the expected performance?
- Is the resource allocation monitoring and the performance measurement sufficiently designed?
- Does the current business function optimally?
- Are synergy possibilities well used?

The second is autonomy versus cohesion. With this element of the variety balance, one asks whether subsystem 1 can make decisions and act independently and thus have a sufficient degree of autonomy. Furthermore, one asks whether the whole system can work strategically cohesively as a whole, although the subsystems have been granted a certain degree of autonomy.

Questions include:

- Is there enough autonomy for the production to do the job?
- Is there enough cohesion to act strategically as a whole?
- Are the vertical and horizontal axes in balance?

1.2.3.4 Variety Balance 4: Change Rate

Variety balance 4 considers the known and unknown future of the whole organisation. The greater the variety of these environments, the greater the variety of system 4 must be. Through this channel, strategic planning becomes the framework for operational planning. However, this often does not work in companies. For this to work, systems 3 and 4 must be equally strong.

Questions include:

- Are we creating new products without a market?
- Do we adopt the changes in the market and technologies?
- Do we only exist through the production of obsolete products?
- Which business units have a slower change rate than the environment?

1.2.3.5 Variety Balance 5: Change vs. Status Quo

Variety balance 5 considers, on the one hand, the organisation's identity and, on the other, looks for an excellent decision-making balance between system 3 and system 4 in order to achieve sufficient strategy development. This means, for example, the proper, timely redirection of human and financial resources from the old to the new business, which is a difficult and, at the same time, important task.

Here the remaining variety is processed, which was not absorbed before.

Questions include:

- Why do we do what we do?
- What do we stand for?

- Does system 5 assume its normative responsibility?
- Is there a good decision-making balance between systems 3 and 4?
- Do we get our strategies implemented in production?

1.3 Modelling with the Viable System Model

In most cases, you will not know the system you are modelling. This means an organisational analysis will be required. This can be done by evaluating existing organisational charts and instructions, processes, reports, manuals, and metrics, the execution of interviews on different hierarchical levels, observation at the place of operation, analysis of meetings, and consideration of the culture. No specific questionnaires are necessary for the interviews; it is enough to talk about the relevant topics. Subsequently, the issues can be matched with the VSM. Experience has shown that workshops about problem topics are well suited, as they provide a lot of implicit information.

In the foreground of graphic modelling is the creative process, not the accurate drawing. The best way is to take a pen and paper and get started. Some model drafts look familiar to a Picasso painting. If you want to reproduce the model digitally, you can use Microsoft PowerPoint, Adobe Illustrator, or any other drawing software.

Important for the understanding here is that VSM is not another representation of an organisation chart: it is a management model that invites us to ask the important (and correct) questions. This goes textually as you can see, for example, with the work of Ivo Velitchkov (2020) and visually.

The following procedure for modelling with the VSM can be recommended.

1.3.1 Modelling Steps

Inside and Now
1. Define system(s) in focus,[7] identify recursion level.[8]
2. Create a system(s) 1.
3. Determine environment and complexity driver.
4. Link system(s) 1 with the environment and complexity driver and identify amplifiers and attenuators.
5. Create S2 and S3*.

---

7 Purpose, tasks, added value.
8 At which level in the system are you?

Outside and Then

6. Create S3, S4, and specific environment and S5.
7. Completeness control.[9]
8. Discussion of the model and refinement.

It is quite possible to jump back and forth between the steps when implementing this procedure. This is normal, and the resulting increase in knowledge is valuable.

1.3.2   Create a VSM Model Using an Example

These modelling steps are illustrated with an example. The example shows a VSM modelling for a construction site from a contractor's point of view.

Example construction site

1. Define system identity:
   - Create a construction project according to the requirements of the client.
   - Make a profit.
2. Create a system(s) 1[10]
   - Construction production is the purpose of the organisation.
   - Construction manager as the local site manager (see Figure 1.9).
3. Determine environment and complexity driver:

Local environment

   - Client
   - Construction supervision
   - Authorities
   - Local authorities and citizens.

Complexity drivers

   - Client
   - Construction supervision
   - Officials (see Figure 1.10).

Figure 1.9   System 1.

---

9 Horizontal system elements and vertical information channels.
10 In this example there is only one S1 defined.

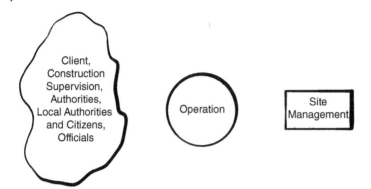

Figure 1.10 Environment and complexity driver.

4. Link system(s) 1 with environment and complexity driver (identify amplifiers and attenuators).

The construction productions consisting of workers, supervisors, and some engineers (e.g. shift engineers) will be connected with the construction management and the local environment. On the one hand, regional interrelations between elements (site management–operation–environment) and, on the other hand, superior interdependencies (site management–environment) are defined. The following attenuation and amplify forces are defined for the capacity of the interrelationships:

- **Attenuators**: pre-sorting, limitations of information tools (filters).
- **Amplifiers**: work instructions or method statements, briefings, transparency, communication (see Figure 1.11).

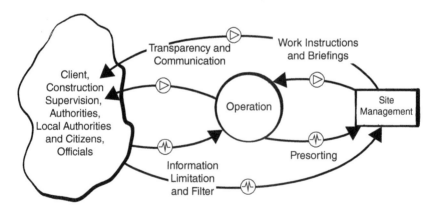

Figure 1.11 Environment and complexity driver connected with retarding and reinforcing forces.

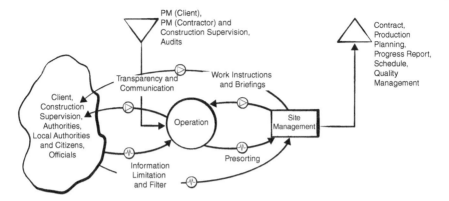

Figure 1.12 S2 and S3*.

5. Create S2 and S3*:

   S2
   - Contract
   - Production planning
   - Progress report
   - Schedule
   - Quality management.

   S3*
   - Project management (client and contractor)
   - Construction supervision
   - Audits (see Figure 1.12).

6. Create S3, S4 and specific environment, and S5:

   The definition for the strategic business of systems 3–5 shall be as set out below.

   S3: Project manager

   S4: Managing director.

   Strategic environment: make a profit, customer satisfaction

   S5: Company (see Figure 1.13)

7. Completeness control (horizontal system elements and vertical information channels):

   After the first draft, it needs to be checked whether all horizontal subsystems and vertical information channels are defined. The model and thus the organisation can be quickly checked for completeness by using the modelling steps as a checklist.

8. Discussion of the model and refinement:

   If the first draft is completed, it is time for discussion. Individual modelling is shaped by one's perspective and needs feedback.

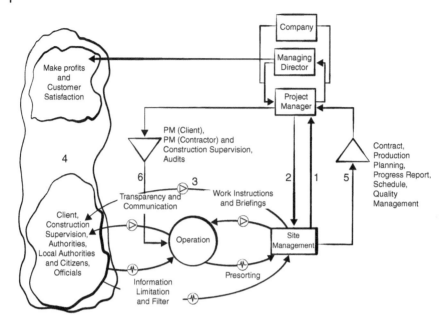

Figure 1.13 S3, S4+ specific environment, S5.

The discussion of the above example has resulted in the following:
- The focus of the analysis is on the operational business of systems 1–3. The analysis of the strategic business of systems 3–5 needs to be intensified.
- Adaption of S4 about strategic environments. Supplement profit and customer satisfaction with customer needs and raw material price development.

Model after discussion (see Figure 1.14)

Analysis of the model

The above model is analysed below as an example. General problems, such as insufficient personnel and the necessity of generating claims for extras, are taken into account.

1. Define system(s) in focus, identify recursion level

   Identify the recursion level. The system in focus is on a construction site.

   Task:
   - Make profits.
   - Generate customer satisfaction for the current project.

   Lever:
   - Manufacture the product according to the customer's order and requirements.
   - To achieve the project and company goals, generate additional services and claims for extras.

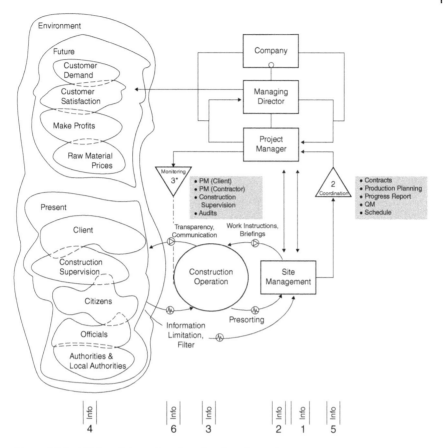

Figure 1.14 VSM model construction site after discussion.

2. Create system 1
   Task:
   - Deliver the construction project efficiently.
   Lever:
   - Use a defined number of resources to achieve maximal output.
3. Determine environment and complexity driver
   Remains blank, since there are no additions to this point from the example
4. Link system(s) 1 with the environment and complexity driver and identify retarding and reinforcing powers
   Task:
   - Management of clients, construction supervision, officials, local authorities, and citizens by using strategic transparency, communication, and public relations.

Lever:

- The correct information must be in the right place at the right time. This can be ensured with the help of suitable centralised and decentralised information systems, which promote self-organisation and better management. A level of autonomy and freedom of decision-making for system 1 compatible with the overall system must be guaranteed.

5. Create S2 and S3*

Task:

- Create a balance between sympathicus and parasympathicus.

Lever:

- Implement communication rules and measurement metrics, promote self-organisation through collaborative production planning, linking work preparation, planning, and execution of production planning.
- Examine external audits, management by walking around, promotion of a good construction site culture and atmosphere for construction workers and employees.

6. Create S3, S4 and specific environment, and S5

Task:

- Project management needs more resources, but the management director does not provide further resources.

Lever:

- Project management must remain persistent and, if necessary, ask for support from the client.

Steps 7 and 8 remain blank, not applicable.

1.4 System Dynamics

System dynamics can be seen as a subscience of systems theory. Jay Forrester developed it in the mid-1950s at the Sloan School of Management at MIT (Forrester 1971). It can be used to identify dynamic, complex, and nonlinear problems and model solutions. Through modelling, the behaviour of systems can be analysed and understood, which ultimately leads to well-founded decisions.

1.4.1 Systemic Archetypes

The Fifth Discipline by Peter Senge (2010), also from MIT, describes five elements that are essential for a learning organisation:

- Team learning
- Building shared vision

- Mental models
- Personal mastery
- System thinking.

Referring to Senge, the unique aspect of a learning organisation is that it is only successful if the disciplines mentioned in this chapter are developed holistically. The interaction and cross-linking of the disciplines are essential since each element can affect the other.

The system archetypes designed by him serve as tools to analyse dynamically complex systems. They describe different behaviour patterns in a system, point out problems, and localise levers to either bring the system back into balance or improve it.

The archetypes are mapped in causal diagrams or flow diagrams. These consist of closed chains of action. The elements in the system are variables that have to reinforce (+) balancing (−) and temporally delayed (‖) effects.

The archetypes are (see Figures 1.15–1.24):

1. **Balancing process with delay** (Figure 1.15)
   Description: Actions are needed to achieve a goal. The organisation acts and assumes that the goal has been achieved. However, the results are not immediately visible, so further efforts are made. Nevertheless, the first action would have reached the goal if a time delay had been included. Too many activities have overshot the goal, and a new unwanted state has arisen in the system.
   Example: A project developer continually develops real estate and has further projects in the pipeline. They implement their projects independently of market forecasts or competitors. However, the market is saturated as demand

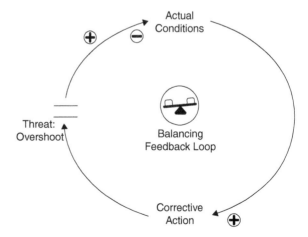

Figure 1.15 System archetype: balancing process with delay.

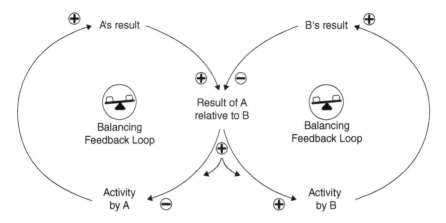

Figure 1.16 System archetype: escalation.

has fallen, and other companies have the same approach (see pig cycle[11]). This leads to unwanted reinforcing feedback, the so-called overshoot.

Lever: Consideration of current market information and the long-term effect of an action (e.g. reducing production capacities). The danger of overshoot is reduced by more cautious dosing of activities.

2. **Escalation** (Figure 1.16)
   Description: Organisation A feels threatened by organisation B and protects itself with a reinforcing activity. Organisation B also feels threatened and also carries out a reinforcing protection action. The entire system swings up to escalation.
   Example: Two companies compete for a job and continually beat each other down in price, reducing their quotations and profits in a noncooperative way.
   Lever: This vicious circle can be ended if the parties adopt a neutral perspective and agree on a common strategy (see also the prisoner's dilemma[12]).

3. **Success to the successful** (Figure 1.17)
   Description: Two people compete for a shared limited resource. One person is more successful initially and can build out their situation by gaining more

---

11 The pig cycle (see also cobweb theorem) describes the problem of the delay in adapting supply to a market. See also 'Overriding': Error in dealing with complexity according to Dietrich Dörner.

12 The prisoner's dilemma models the situation of two prisoners. Both prisoners are interrogated individually and cannot communicate with each other. If both deny the crime, both receive a low penalty. If both confess, both receive a high penalty. If only one reveals, that person will earn impunity from prosecution, while the other will receive the maximum punishment. The dilemma is that each prisoner must decide either to deny (i.e. cooperate with the other prisoner) or confess (i.e. betray the other prisoner) without knowing the other prisoner's decision.

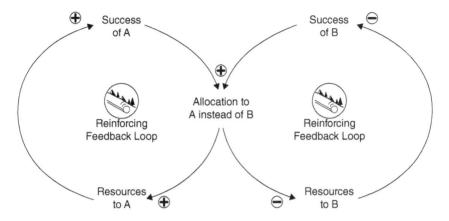

Figure 1.17 System archetype: success to the successful.

access to the resource as a 'successful person'.[13] The other person, in turn, gains less access to the resource until it is ultimately no longer available.

Example: Two departments of a construction company both develop a smartphone app to make everyday work more efficient. For the implementation, they need support from the IT department. Department A invests a lot of time in the beginning and, compared to department B, can present a complete concept with promising prototypes at a very early stage. The IT department can extract all relevant information from the idea, and the programmers begin their work. Further information is constantly needed from the other department so that the programmers cannot start and the IT department uses all its existing resources for department A.

Lever: The solution to the problem is to find in a common overarching goal and the connection of the activities.

4. **Limits to Growth** (Figure 1.18)
   Description: The growth of a process increases continually to a point where growth begins to stagnate. The limited capacity of a resource causes the slowdown. To achieve any growth now, more significant effort or expense is required.
   Example: A start-up grows continually to a point where it can no longer progress with existing structures and competencies. More professional management skills and other organisational structures are needed to ensure further growth.

---

13 See also 'Choices with externalities', which means that an individuum must consider the choice of coexisting individuals.

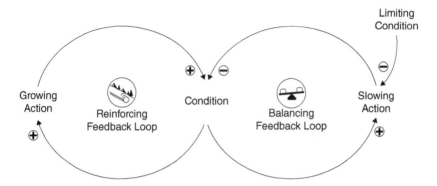

Figure 1.18 System archetype: limits to growth.

Lever: If the start-up becomes too large, new start-ups from the original organisation (keyword: holacracy[14]) must be founded to maintain its unique culture and dynamics.

Growth is good, but only to the proper extent, the dependency is to be avoided; this contradicts the principle of self-regulation. Two guiding principles: (i) if a system wants to grow and survive simultaneously, it must undergo metamorphosis and (ii) first it has to be deficient and then it has to develop systematically.

5. **Shifting the burden** (Figure 1.19)
   Description: An existing problem has several symptoms. The symptoms are treated for rapid success. The actual problem is not considered.
   Example: A project manager fails in reducing the stress of the project business through relaxation techniques. To calm down as a quick fix, they start smoking. Over time, they reach their stress limit earlier and so smoke more often. Owing to the new problem, i.e. nicotine addiction, they gradually lose their natural ability to relax.
   Lever: Understand the symptomatic solution as the possibility of bridging. Use the resources to solve the fundamental problem, even if no improvements are discernible for the time being.

6. **Fixes that fail** (Figure 1.20)
   Description: An assumedly correct measure to solve a problem has unexpected long-term consequences that reinforce the original problem or creates a new one.

---

14 Holacracy: This turns away from traditional management styles and distributes responsibility among employees: everyone is encouraged to make decisions and thus becomes a co-entrepreneur. The following principles are applied: self-organisation, cybernetics, agile methods, and collective intelligence.

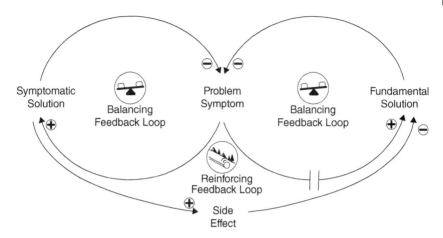

Figure 1.19 System archetype: shifting the burden.

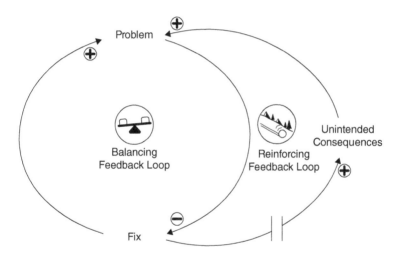

Figure 1.20 System archetype: fixes that fail.

Example: A construction company has a good order backlog and needs suitable personnel to process it. Owing to a shortage of personnel, a project manager from ground engineering is appointed for building construction. The problem appears to have been solved, but the project manager lacks the building construction expertise to manage the project effectively. Wrong decisions are made, which leads to additional costs.

Lever: The short-term corrective action shall be applied only until a long-term solution is found.

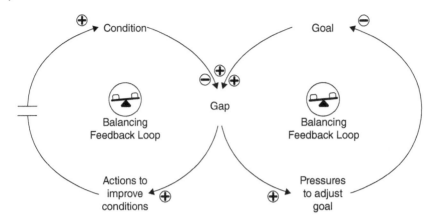

Figure 1.21 System archetype: eroding goals.

7. **Eroding goals** (Figure 1.21)
   Description: There is a gap between the target state and the actual state. To achieve a short-term goal, the target position is downgraded so that another long-term goal cannot be achieved.
   Example: Owing to a delayed performance by construction suppliers, the construction work as defined at the start of the project could not be achieved. The project manager adjusts the target values to communicate that success and the daily output were achieved. By reducing the target output, however, the project's total time is extended, and the original completion date is not reached.
   Lever: The long-term goal must be kept in focus. Achieve both short-term and long-term goals by increasing resources.

8. **Accidental Adversaries** (Figure 1.22)
   Description: Two individuals agree to cooperate to achieve a goal. However, the goals cannot be combined, so that the cooperation does not bring any advantages.
   Example: Two companies form a joint venture to win a significant construction project tender. However, the interests of the two companies are so different that the companies inhibit each other and do not achieve the actual goal.
   Lever: Cooperation can succeed if the mutual objectives are either complementary or identical. A possible cooperation interface for the counterpart through intensive dialogue should be found.

9. **Tragedy of the commons** (Figure 1.23)
   Description: The individuals in the system use a resource to their advantage. Each person uses it regardless of the fact that the other parties in the

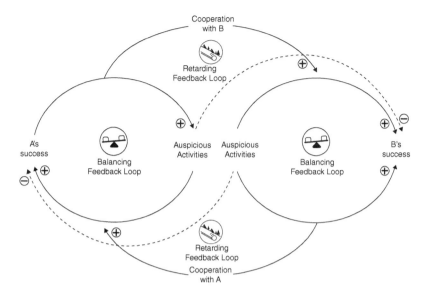

Figure 1.22 System archetype: accidental adversaries.

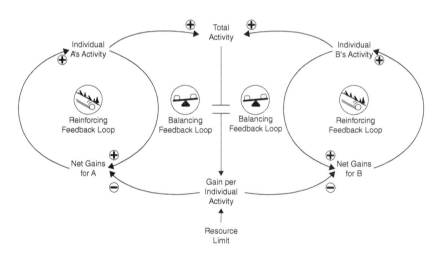

Figure 1.23 System archetype: tragedy of the commons.

system use the same resource. Through collective use, the resource is used up in the end.

Example: Two managers of a company both need a valuable employee for their work who is accessible to both. The employee achieves good results, from which both executives benefit. However, they do not take into account the employee's capacity to work, so that after a certain period of time the employee is overburdened and no longer able to work.

Lever: The problem can be solved by looking at the entire system and not just at one's own view. Through a sensitive and objective consideration of the actions in the system, resource usage should be sustainable.

10. **Growth and underinvestment** (Figure 1.24)
    Description: If growth starts to stagnate, the organisation's capacities can be increased by investments. The targets are scaled back to justify underinvestment if no investments are made.
    Example: A company grows continually and would have to train its employees to have suitable personnel for the growth phase. However, the company decides not to do so and continues to offer the same service without further developing quality. The customers are no longer satisfied with the service in the long run and change the company.
    Lever: Forward-looking planning and investments enable continual growth. If there is an opportunity to grow, capacities must be expanded.

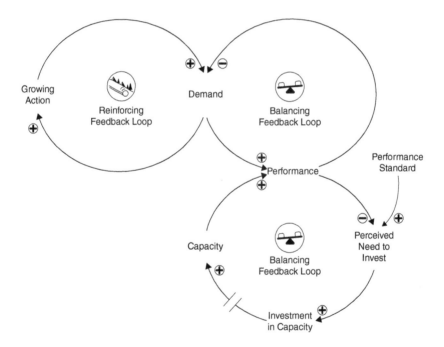

Figure 1.24 System archetype: growth and underinvestment.

1.4.2 Modelling with System Dynamics

Modelling with system dynamics has a more formal mathematical background than other system approaches. Qualitative and quantitative models analyse and design linear and nonlinear economic and business situations.

- Qualitative models (causal diagrams): These models can be created relatively quickly and show dependencies and unknown causalities without hard data, e.g. using the influence or knowledge matrix. Quantitative analysis should follow a qualitative analysis.
- Quantitative models (flow diagrams): With these models, all processes can be mapped mathematically and with actual numerical values. It can be modelled as a function of time. The result is a self-contained, mathematically exact result.

In practice, the method is applied similarly to statistics in strategic and operative corporate planning and the concretisation of balanced scorecards or for model-based learning. Scenario analyses support the decision-making process. Relevant programs are suitable for conception, creation, and modelling.[15]

1.4.3 Example: Managing Risks with System Dynamics

At University College London and the University of Leeds, system disciplines are merged with project management in the context of large projects. Giorgio Locatelli professor at the University of Leeds, sees many advantages in using systems engineering[16] to manage the existing complexity in large and megaprojects (Locatelli et al. 2014).

According to Miller and Lessard (2001, 2007), risks occur over the entire project duration and continue to develop. Threats can reinforce each other, but they can also balance out. The interaction of risks over a certain period can be analysed with system dynamics. Prince Boateng, a researcher at Robert Gordon University in Scotland, recognised the potential of system dynamics in large and megaprojects and used system theoretical approaches to investigate the risks in a megaproject in Edinburgh (Boateng et al. 2012).[17] The result of system dynamics analysis is remarkably accurate. The methodology of Boateng (2012–2016) is explained below. The procedure includes two different phases.

---

15 See also: Consideo, AnyLogic, iThink/Stella, Dynamo, Vensim or Powersim.
16 System engineering integrates engineering disciplines with management to develop and control complex technical systems. System engineering uses various tools for decision-making. One example is the design of models and simulations for analysis using system dynamic models.
17 Boateng developed the model in 2012 during the construction phase. The project was completed in 2014.

Boateng used an analytical network process (ANP) to prioritise the most critical risks as a first step. For this purpose, potential risks were listed and designated as options. These options were categorised using the STEEP method (Design Methods Finder 2022). Subsequently, the impact of the options on cost, schedule, and quality was considered. The prioritisation was based on statements from experts.

The second phase included the development of the system dynamic model. Within this phase, a qualitative causal diagram was created. Subsequently, a quantitative analysis using a flowchart was performed.

The system dynamics model predicted a £270 million cost overrun and a delay of 2.15-years. The actual cost overrun was £230 million and the time delay was three years. This represents a cost accuracy of 85% and a time accuracy of −40%.

Depending on variables and parameters, a model can be more or less meaningful. The challenge is to define realistic initial values. Even without complex quantitative modelling, even a simple causal diagram supports understanding the dependency and dynamics of risks.

1.5 Findings, Criticism, and Reflective Questions

1.5.1 Findings

As a rule of thumb, one can say: complicated is about the composition of a system. If there is enough time and competence, the system can be understood. Complexity is about the number of elements, their relationships, and the variability of the system states over time.

Complexity in the construction industry is driven by institutions, actors, time and cost pressure, interfaces, uniqueness, long preparation and construction time, changes in standards, growing critical interest, and know-how loss among commercial personnel.

Chaos describes the noncausal relationship between cause and effect, especially if small changes in the initial conditions significantly impact the output (keyword: nonlinearity).

The Cynefin framework categorises systems as simple, complicated, complex, and chaotic.

Beer's (1974) VSM is a reference and analysis model for organisations. Organisational units should have similar generic organisational structures at different hierarchical levels. Keep in mind the static and the dynamic perspective on the model. According to Hoverstadt (2019): 'Go to where the chaos is and bring order.'

Referring to Ashby (1985), social structures remain manageable if the steering system (management) has the same variety or the same variety of actions (possibilities) as the system to be managed (team + environment) itself. According to

Luhmann (1994), an environment that is tolerable for the system has to be selected for this.

System dynamics serves the holistic analysis and simulation of complex and dynamic systems.

Through modelling, the behaviour of the systems can be analysed and understood, which ultimately leads to well-founded decisions.

Remember Senge's (2010) system archetypes:

- Balancing process with delay
- Escalation
- Success the successful
- Limits of growth
- Shifting the burden
- Fixes that fail
- Eroding goals
- Accidental adversaries
- The tragedy of the commons
- Growth and underinvestment.

These describe different behaviour patterns in a system, identify problems, and locate levers to either bring the system back into balance or improve it.

The analysis and design are carried out using qualitative models (causal diagrams) and quantitative models (flow diagrams).

1.5.2 Criticism

The entire complexity approach usually does not provide quick and easy answers to complex questions. This means that one has to deal with questions in-depth, especially with the methods to define questions and generate solutions.

According to Mike C. Jackson (2003) , a former professor of management from Hull University Business School: 'Any model is, necessarily, a partial representation of reality', so is the VSM. According to Jackson (2003), answers to the human component about individual purpose and motivation are missing in the model, like answers about participation and democracy. The model can become more than a sophisticated control system in the wrong hands.

The authors observe further weaknesses of the model. At first glance, it is abstract and too rich to quickly generate enthusiasm unlike other fashionable approaches. The VSM representation, which resembles a TV or radio circuit diagram, does not reveal the dynamic dimension of the model. People often think it is a different representation of an organisation chart, which is not as described.

With the system dynamics method, everything can be modelled quantitatively and qualitatively using feedback loops. The knowledge of archetypes provides a better understanding of the system. The advantages of the approach are also disadvantages at the same time.

The models can suffer from inaccuracy because they may ignore approaches from the relevant disciplines and may be unsuitable as investigative tools. As with any mathematical modelling, attention must be paid to the model or equation, the data, and the calculation process. The calculation and the results can very quickly move away from the practice and the practical question.

1.5.3 Reflective Questions

The reflective questions asked in this chapter are recapped here.

- Is it more sensible to increase the variety of our system, or should we change our environment?
- What are the complexity drivers of our system or organisation?
- Is the suitable method or the right mix of techniques applied to solve the tasks? Keep the Cynefin framework in mind!
- Does the organisation create stability internally and externally? Is the horizontal and vertical variety in balance? What needs to be done to create stability?
- In addition to the elements, what are the relationships between the elements? They say a lot about the system's behaviour. (Who does what with whom?)
- The production is the reason the organisation exists. Did the organisation pay a high level to the production system? Did the organisation follow the production?
- Is there enough coordination to work efficiently and effectively in the production system?
- Is there a good performance/resource ratio? Are there enough resources to perform well? Is the given resource used well?
- Is there enough autonomy for the production to do the job? Or is there too much micro-controlling?
- Is there enough cohesion to act strategically as a whole?

2

Lean Management and Lean Construction

We have modified our environment so radically that we must now modify ourselves to exist in this new environment.

Norbert Wiener

Designing Intelligent Construction Projects, First Edition. Michael Frahm and Carola Roll.
© 2022 John Wiley & Sons Ltd. Published 2022 by John Wiley & Sons Ltd.

2.1   Pioneers of Lean Management

Most consideration in history of the development of lean management focuses on its most prominent pioneers: Henry Ford and the Toyoda family, but also William Edwards Deming, Joseph M. Juran, Masaaki Imai, Taiichi Ohno, Shigeo Shingo, and Kaoru Ishikawa (Brunner 2011; Gorecki and Pautsch 2013).

However, a closer look reveals that the first essential approaches, later reflected in the lean philosophy, go back further historically. These include pioneering important innovations that gave rise to new technologies, even if these did not yet lead to the widespread application due to the still weakly developed industrialisation. Some examples of this are shown in this chapter.

For example, as early as the late fifteenth century, galleys were being built at the Arsenale Novissimo, the Venetian shipyard, using the principle of flow-forming. Moreover, merchant ships were designed to be converted into warships comparatively quickly (Lean Enterprise Institute 2007).

Beginning in 1784, US inventor Oliver Evans developed an automatic grain mill invention. By using conveyor mechanisms arranged in rows and monitoring the milling result, the production steps from raw grain to finished milled flour could be carried out fully automatically for the first time, resulting in a more uniform milling result as well as a higher level of hygiene in the handling of the product (Jacobson and Roucek 1959).

The US inventor Eli Whitney Jr. is generally described as the founder of interchangeable construction since he developed the principle of the standardised production of interchangeable parts to manufacture weapons around 1790. However, there are indications in the literature, including Woodbury, Battison, and Ferguson, that there are justified doubts about this attribution (Hounshell 1984).

The industrial use of assembly-line production finally became widespread in the first half of the nineteenth century. For example, in 1833, an assembly line was used in England to produce ships' rusks. From 1870, the slaughterhouses of Cincinnati used a conveyor technology in which the slaughtered animals were transported from worker to worker on a circular chain ('disassembly lines'). This application was eventually perfected in the Chicago stockyards, where the individual disassembly steps were distributed among various workers, resulting in massive time savings. For example, according to Dominic Pacyga, a professor at Columbia College in Chicago, 'around 1890 […] it took an experienced butcher and his assistant eight hours to slaughter and cut up a cow. In the stockyards it took only 15 minutes' (Pretting 2006).

These disassembly lines inspired Henry Ford to introduce his assembly lines in Detroit in the spring of 1913. Here, it is documented that, by saving the workers' walk, the cycle time could be reduced from 2.3 to 1.19 minutes. However, it must be noted that the use of the assembly line could not be realised until parts were easily interchangeable. In combination with the reduction to only one step per worker, Ford reduced the time per work cycle, which had previously been 514 minutes, to 2.3 minutes (Womack et al. 1992).

2.2 Toyota Production System and Tools

Lean management can be traced back to a large extent to the Toyota Production System (TPS), which in turn was formed from numerous formative influences. Gorecki and Pautsch characterise lean as a philosophy significantly shaped by the flow principle as applied at Ford, the Jidoka principle of Kiichiro Toyoda, kaizen, the PDCA cycle according to Deming, as well as the involvement of Taiichi Ohno (Gorecki and Pautsch 2013).

The TPS consists of two core columns which were founded by Sakichi and Kiichiro Toyoda (Toyota Material Handling 2021):

- Just-in-time (JIT) production, which means producing only according to customer requirements.
- Jidoka, which means creating transparency in the production process.

Ohno designed the TPS by refining the core columns and adding other essential elements, namely heijunka, kaizen, and standardisation (see Figure 2.1), to increase added-value activities and eliminate waste. 'Added value' is defined by the actions for which the customer is willing to pay. You need to know the customer's needs and your production processes for added value. The value stream must be understood.[1] The right thing must be delivered at the right time – and always from the perspective of the internal or external customer.

These include:

- To bring process steps into an optimal sequence.
- Synchronise process steps with each other (JIT or parallel).
- Create continuous flow ('one piece flow' instead of processing in large batches).
- Balance work content and remove bottlenecks.
- Level out fluctuations in demand or orders.

Brunner goes even further here and describes TPS as a 'comprehensive technology management program' based on seven pillars (kanban, heijunka, 5S and 5W,[2] jidoka, kaizen, LPM, and One-Piece-Flow cells) (Brunner 2011).

---

1 The value stream can be displayed transparently and uniformly using value stream mapping (value stream mapping = actual state). The value stream analysis is the basis for process optimisation (value stream design = target state).
2 5S: seiri = sort, seiton = set in order, seiso = shine, seiketsu = standardise, shitsuke = sustain/ self-discipline. 5W, or the 5-why-method, is a method in the field of quality management for cause/effect determination. The goal of this application of the five 'Why?' questions is to determine a cause for a defect or problem. LPM: lean production management.

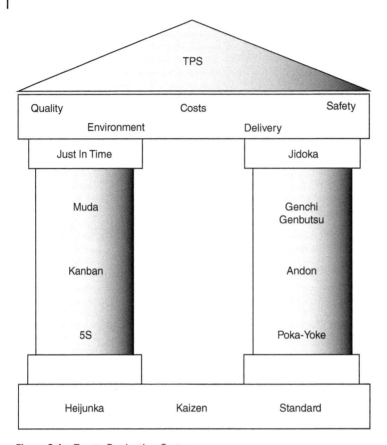

Figure 2.1 Toyota Production System.

The reasons for the development of the TPS were the economic conditions in Japan after the Second World War, the unique features of the market there, and the intense competitive pressure from mass production in the automotive industry, which was particularly prevalent in the USA. Mainly due to the scarcity of raw materials and the need for small quantities with a high number of variants and high-quality requirements at the same time, Toyota's production system differed significantly from Ford's system.

Shingo identifies two significant differences: Toyota concentrated on producing small batches using the principle of mixed-model production, whereby different vehicle types are manufactured on one production line. In addition, all work processes were consistently designed for batch size 1 (Shingo 1992). These are the most obvious distinguishing features, but further examples can be found in the literature. Parkes, for example, referring to Taiichi Ohno and Bodek's work *Lean Management: Beyond Large-Scale Production* (1988), also mentions the use of JIT and kanban, i.e. pull-oriented processes in contrast to Ford's conveyor belts, the support of automation by human labour, the substantial reduction of inventories, as well as the avoidance of the production of overhangs and rejects (Parkes 2015).

In addition, there is a clear difference between the two production systems in the way workers are treated. At Ford, the work processes were strongly influenced by the teachings of Fredrick Winslow Taylor, which understood people as a production factor and led to the alienation of workers from their task due to the monotony of the very dissected work steps (Taylor 2013 cited in Wirtschaftslexikon24 2018).

The TPS considers employee involvement as one of its focal points, according to which employees represent the company's greatest asset (Brunner 2011).

Even though the roots of lean go back to the middle of the last century, the objective of the lean philosophy has lost none of its relevance. For example, today's displacement markets have similar conditions to those once found in Japan after the Second World War (Gorecki and Pautsch 2013). High raw material costs, the desire for individualisation by customers, and increased global market transparency due to worldwide networking do the rest. Bertagnolli also points to the importance of lean for new methods such as Scrum or as a basis for optimising processes before their intended digitalisation (Bertagnolli 2018).

The system, which Toyota developed, has gained international recognition through a study by the Massachusetts Institute of Technology (MIT) (Womack et al. 1990). In this study, the researchers coined the term 'lean production' for the TPS. 'Lean' is intended to clarify that the production system does not incorporate any other unnecessary processes into the production processes. According to the DIN EN ISO 9000 standard for quality management systems (Deutsches Institut für Normung e.V. 2015), a process is generally defined as: 'a set of interrelated and interacting activities that converts inputs into results'.

Lean production has attracted the attention of many different people. From this, universal and sector-independent use of this approach has developed. Consequently, the variety of applications of lean philosophies is unlimited.

2.2.1 Waste, Kanban, and Just-in-time Principle

The avoidance of waste is one of the most major requirements in lean management. Different levels of waste (Japanese: *muda*) are identified. Thus, in addition to the work that adds value, Takeda describes waste in terms of machinery and equipment (*kanji muda*), work processes that represent waste but are unavoidable (*hiragana muda*), and everything that represents absolute waste (*katakana muda*) (Takeda 1995).

The eight typical types of waste in lean management are defined as:

- **T**ransportation
- **I**nventory (stocks of products, materials, consumables, etc.)

- Unnecessary **M**ovement
- **W**aiting
- **O**verproduction (produce more than needed)
- **O**verprocessing (unnecessary work steps, oversized plants, etc.)
- **D**efects and rework
- **S**kills (not using employee's full abilities).

Mnemonic: **TIM WOODS** – initial letters of the eight typical types of waste.

Older publications usually list only seven types of waste ('TIM WOOD'). More recent publications also include the unused creativity potential of employees among the types of waste: skills (Brunner 2011).

A classic example of waste in construction projects is the definition of oversized buffers by preliminary or subsequent trades. A further example is the culture of conflict and its associated supplementary management, which diverts the focus from the actual production. Another obvious indication of waste on construction sites is how well organised they or their equipment areas are. How well a site is organised, how tidy and logically set out it is says a lot about its productivity, work safety, and management.

In addition to the unconditional avoidance of waste, the TPS contains numerous approaches that control and optimise the flow of materials as well as all other value and information flows.

Based on the concept of American supermarkets, where the shelves were refilled according to demand, Ohno invented kanban cards to ensure that the right components for production were available (Ohno 2013).

Kanban cards are a highly visible tool, which is used in the TPS for the component requirement as and when needed. This means that only a minimum stock of components is kept in the assembly area. Before stocks run out, a kanban card with instructions from the operator ensures JIT delivery (see Figure 2.2).

As an alternative to kanban cards, these control loops can also be controlled via the containers required for transport. For this purpose, the information required for identification and control loop management is attached to each transport container itself. The material flow is controlled by monitoring and the timely replacement of used containers with newly filled ones. This means that when an empty container is received at the source an order for material refilling is triggered. When selecting the transport containers, it is necessary to ensure the size, shape, and planned number of contained parts, as well as the handling, safety, and distinguishability of the containers.

The described supermarket pull system represents an essential element of JIT production.

Just-in-time production (abbreviated to JIT) means delivery 'just at the right time'. Alternatively, the technical term 'demand-synchronous production' has also become established (Kamiske and Brauer 2011).

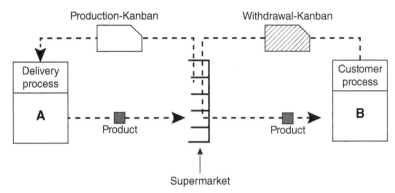

1. Customer Process goes to the supermarket and takes out what is needed and when it is needed.
2. Delivery Process produces, to replenish what has been withdrawn.

Figure 2.2 Supermarket pull system according to Rother and Shook (2011).

The JIT principle balances and levels the production programmes, the pull principle, the flow production, and the takt time (Toyota Motor Corporation 1998).

This concept, which Taiichi Ohno developed as a central part of the TPS, involves a complex and decentralised organisational and control concept. Only the material needed to fulfil the individual customer order is delivered at precisely predetermined times during production.

To achieve this, Ohno wanted to merge many suppliers and parts factories into one large machine, producing only as many units as the downstream production stage required. This, in turn, meant that any inventory was eliminated, making the system extremely vulnerable to even the slightest disruption. From Ohno's perspective, however, this was precisely the decisive advantage of the system: the absence of all buffer-induced safety stocks sensitised the attention of everyone involved in the production process to think and act with foresight and rectify any problems before they led to malfunctions (Womack et al. 1992).

This is implemented by ordering the required materials or prefabricated components precisely in line with the final assembly requirements. A vital role is played here by the close involvement of suppliers, who are obliged to deliver the ordered materials directly to the processing site or in the immediate vicinity within a defined lead time. Subsequently, the provided materials and components are usually installed immediately; with JIT, there is no need for warehousing or a subsequent picking process upstream of production.

Owing to the close timing of material supply and final assembly, JIT production is highly susceptible to external disruptions (production-related, traffic-related, and meteorological). To reduce the risks of production interruptions resulting from this, numerous companies using the JIT principle have therefore implemented a risk management system tailored to this. In addition to the establishment of alternative suppliers by the customer, the risk-reducing measures also include the application of high contractual penalties to suppliers in the event of late delivery. This leads to suppliers locating themselves in the immediate vicinity of the customer or directly on the customer's premises.

By consistently implementing the JIT principle, the manufacturer no longer needs to maintain storage capacity for input materials. As a result, this process has become established in the automotive and aircraft industries, where large numbers of parts and assemblies with many variants are procured from external suppliers and delivered directly to the assembly line.

The further development of JIT is just-in-sequence (JIS) delivery. Here, the required material is ordered and delivered precisely in the sequence in which a product is installed, as shown in Figure 2.3 (Herlyn 2012).

2.2.2 Jidoka and Related Elements

In addition to JIT production, the jidoka theme circle (including genchi genbutsu, andon, and poka-yoke) represents the second pillar of the TPS.

In the term *Jidōka*, Taiichi Ohno, who coined the term, combined the Japanese word for 'automation' with the Japanese character for 'human being' to express that, in the context of machine-based production processes, the possibility of integrating humans as a supervisory instance should also always be taken into account (Furukawa-Caspary 2016).

Therefore, the term is also translated as 'autonomous automation', 'intelligent automation', or 'automation with a human touch', among others (Toyota Motor Corporation 2021).

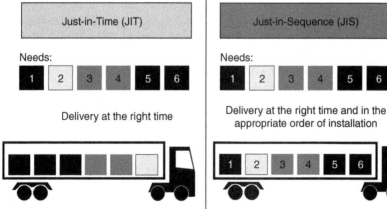

Figure 2.3 Difference of JIT and JIS.

The principle is based on the invention of the self-acting loom by Sakichi Toyoda, the founder of the Toyota Motor Company. With this, breaking a warp or weft thread led to the automatic stoppage of the loom, which effectively prevented further defective production (Syska 2006).

The transparency established through jidoka promotes the increase in the quality of the products. Quality controls are introduced in each subprocess to ensure the early detection of defects. Each team member is responsible for carrying out quality controls before the intermediate product is passed on to the next station. When a defect is identified, a solution is immediately sought. At Toyota, for example, this means that as soon as anomalies are detected the machines stop by themselves. The worker in charge then stops the production line to find out the fault (Rosenthal 2002).

Thus, jidoka represents a part of the quality control process. For this purpose, the following steps are carried out:

1) Detection of the anomaly.
2) Machine stop.
3) Problem-solving.
4) Detection of the root cause and implementation of countermeasures.

Thus, the primary goals of jidoka are to prevent production errors and create a basic understanding of problems that may occur to prevent them in the future.

Genchi genbutsu is one of the tools within the framework of the jidoka principle and means going to the source or cause and assessing the problem yourself to get a complete and correct picture so that a solution can be found (Gorecki and Pautsch 2013).

The focus here is on the observer's requirement to go directly to the place of action, or value creation, where any decisions will have an impact. In this way, the manager should also become increasingly aware of their role as a coach for the employees within the framework of kaizen.

An essential term in this context is 'gemba'.

'Gemba' is generally understood to mean the place where value creation takes place. The central idea of lean in this respect is that those responsible do not allow themselves to be reported to when problems arise, but instead form their picture of what is happening on site, as this makes it much easier to uncover grievances; in addition, employees on site represent a valuable source of information (Gorecki and Pautsch 2013). To make the most of this effect, it is advisable to follow gemba's six principles:

1) Go to gemba (the place of manufacture) when a deviation occurs.
2) Check genbutsu (employees and machines in the manufacturing process).
3) Take immediate action.
4) Find out the cause of the deviation.

5) Eliminate the cause at its source.
6) Standardise to prevent recurrence of deviations (Zollondz 2001).

The andon is used to quickly and easily detect possible quality problems, machine downtimes, or other deviations in the 'gemba walk', which represents the visualisation of the jidoka principle (Gorecki and Pautsch 2013).

The 'andon system' is mostly implemented with information boards: the 'andon board' is a simple but highly visible panel that shows the status of the production lines. It immediately informs management when an employee notices an error and precisely identifies where the error occurred.

In addition to the andon board, which today is mostly in digitalised form, optical inspection can also be carried out by means of light signals (andon light) on the respective machines or production lines (see Figure 2.4). In general, the andon system is structured according to the following six-stage problem-solving cycle:

1) Recognition of the problem.
2) Andon signal.
3) Evaluation of the problem.
4) Limitation of the effects.

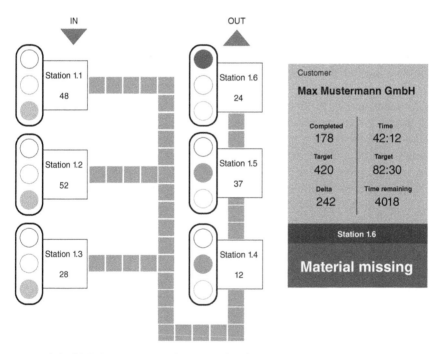

Figure 2.4 Digital andon board of a production line.

5) Analysis of the causes.
6) Avoidance in the future.

The ultimate goal of the TPS is to eliminate all waste, which requires a stringent pursuit of the zero-defect goal. To achieve this, a continuous cycle of process review and improvement is required, called baka-yoke.

A part of this process is a simple, but extremely effective, implementation of sustainable defect prevention: 'poka-yoke', a Japanese term that stands for 'defect-checking devices' (literally 'avoid unfortunate mistakes') (Shingo 1992).

Poka-yoke ensures that the right conditions are in place before a process step is executed, thus preventing the occurrence of errors from the outset. Here, the poka-yoke system matrix in Figure 2.5 shows a possible solution path (from left to right).

There are countless examples of poka-yoke in production but also everyday life. One example is the introduction of USB-C cables, which are easy to connect, and there is no need to check in advance which side of the cable has to be inserted at the top or bottom. Another example is connectors with a defined direction (e.g. for reverse polarity protection). These are geometrically designed so that incorrect plugging (which could damage the respective device) is prevented from the outset, as shown in Figure 2.6.

Reasonable poka-yoke solutions are characterised by the following features:

- Require only a small investment.
- Can be implemented quickly and easily.
- Have a significant positive impact on product quality.

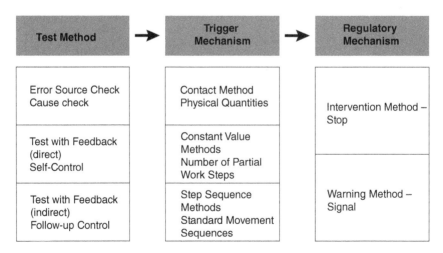

Figure 2.5 Poka-yoke system Matrix according to Bergbauer (2008).

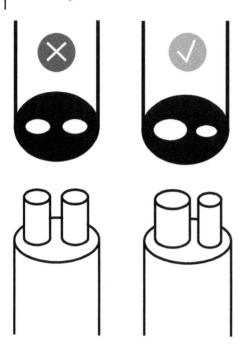

Figure 2.6 Possible implementation of poka-yoke.

- Are part of the process and do not represent an additional work step.
- Are developed together with the employees concerned.
- Support high-quality work instead of control.
- Final inspection can be omitted.

2.2.3 Heijunka

The following features characterise the reasonable poka-yoke solutions concept of heijunka as one of the prerequisites to design JIT processes. 'Heijunka' means the harmonisation ('smoothing and levelling') of the production plan regarding quantities and varieties.

For this purpose, the heijunka board is used as a tool, representing the physical implementation of the levelling idea and is intended to enable demand-oriented production (see Figure 2.7). For this purpose, the board is to be subdivided with a grid that maps at least 24 hours and contains all associated part numbers or assembly lines. Levelling kanban cards are then inserted into the grid accordingly to coordinate the flow of materials through production. This makes the heijunka board a control element for intralogistics and a clock generator for production.

Owing to the visual structure of the heijunka board, production planning for an entire week can thus be overviewed, and the sequence of daily call-offs and their

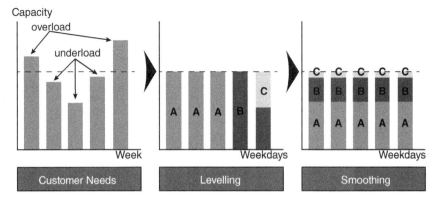

Figure 2.7 Example of levelling and smoothing according to VDMA Forum Industrie 4.0 & PTW Institut für Produktionsmanagement, Technologie und Werkzeugmaschinen (2018).

degree of fulfilment can be measured. This system is completed with tools that enable quick and effective intervention when process problems occur (Gorecki and Pautsch 2013).

Since the demand for the levelling and smoothing of production is often challenging to implement without deviations, in reality Rother (2013) recommends interpreting heijunka as a pattern or optimal target state.

For this purpose, Rother (2013) recommends the following steps:

1) Load the heijunka box/sorter according to the planned sequence and maximum batch size.
2) Ask, 'Can we produce like this today?'
3) Yes – start production. No – 'What is stopping us?'
4) Detect problem, fix it, and return to a planned sequence as soon as possible.

Furthermore, Heijunka plays a vital role in dealing with mura and muri. 'Mura' describes all adverse effects from a nonsynchronised overall sequence of individual subprocesses. The consequences of this are, for example, buffers and stockpiles to absorb and compensate for uncertainties in a process. A lack of harmonisation of the subprocesses leads to adverse effects on the flow in production and to a suboptimal production flow.

'Muri' describes the overload of individual parts of the production process (Verlag moderne Industrie GmbH 2017). Overloading can occur in machines as well as in the human body. In the case of machines, it causes increased wear and tear; in humans, it leads to a high level of stress or sick leave.

Muri and mura can be identified using a value stream analysis in the overall process. Overall, it can be concluded that muri is the cause of mura, and mura leads to muda.

2.2.4 Single Minute Exchange of Die (SMED)

Single minute exchange of die (SMED) can be translated as a tool change in the single-digit minute range. The process used here, which is intended to reduce setup times and thus inventories in front of the machines, was invented by Shigeo Shingo. For the procedure, eight principles have emerged since the 1970s. The central techniques here are:

- Distinction between internal and external makeready processes.
- Transfer of internal and external makeready processes.
- Optimisation and standardisation of internal and external makeready processes.
- Elimination of adjustment processes.
- Parallelisation of setup processes.

Among other things, it is essential to keep internal makeready times as short as possible (also at the expense of external makeready times). Furthermore, practical, quickly interchangeable fastening methods are to be developed and used and, lastly, stepless adjustability of fixtures and machines is to be avoided as far as possible (Shingo 1992).

2.2.5 Kaizen and Standards

The term 'kaizen' means change for the better or continuous improvement. It is the basis for a steady improvement of processes by looking for innovative solutions and other added values for the customer. In particular, the ideas of employees who are directly and indirectly involved in the value-added process are decisive for value-creating optimisation.

Employees make suggestions for improvements to all work and company processes by a standardised procedure, either through self-motivation or at the request of their supervisor.

After the suggestions have been submitted to the work team, they are discussed, accepted if necessary, and implemented according to the PDCA cycle. In the case of more far-reaching changes, additional people (from other hierarchy levels) may be called in. It is essential that the submitter has the right to feedback and that accepted suggestions are implemented quickly (Gorecki and Pautsch 2013).

While in Japan, kaizen is seen both as a general philosophy of life and as a general philosophy of work. As a methodical concept for striving for continuous improvement (see Figure 2.8), in the West, its meaning is often reduced to 'continuous improvement'.

It should be noted that the core requirement of kaizen is mainly the constant questioning of basic assumptions (Furukawa-Caspary 2016).

So it is essential to set no limits to employees' creativity and be open to innovative solutions. It is somewhat secondary whether a small or large kaizen has been

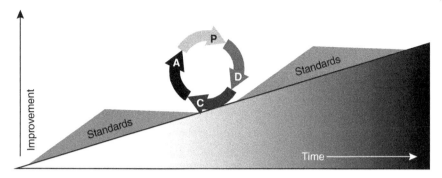

Figure 2.8   Relation between kaizen, PDCA, and standards.

introduced. What is more important is constant and continuous improvement. If the Toyota philosophy is fully realised, the phase in which kaizen is implemented is ongoing and never finished.

The objective is the constant development of the company's level of innovation and the prevention of operational blindness.

Owing to the more generalised understanding of kaizen in Japan, many tools and methods, too numerous to mention here, are included in the kaizen topic complex.

To understand the original objective (or original intention) of kaizen, it is important to have a look at the five central bases of kaizen:

- process orientation
- customer orientation
- quality orientation
- criticism orientation
- standardisation.

As can be seen from this list, standardisation is one of the essential foundations of kaizen, or in short, 'Without standardisation, there is no kaizen' (Gorecki and Pautsch 2013).

Standardisation is elementary for detailed production planning and crucial for quality assurance. Standards reduce the variety in the production process and therefore decrease complexity. Standardised workflows and parts ensure a high level of quality and stable processes and allow for high scalability.

2.3   Lean Management and Its Principles

The term 'lean management' was first coined by James P. Womack, Daniel T. Jones, and David Roos among others. The MIT researchers used it to describe the observations they made during a benchmark analysis. The research project revealed the significant differences between the production systems of primarily Western and Asian automakers.

However, Weis also points out in his definition that 'lean' encompasses more than mere streamlining. Instead, lean is a philosophy that applies diverse principles, measures, and methods to planning, designing, and controlling the value chain for both goods and services (Weis 2013). Gorecki and Pautsch further specify this approach. For example, they cite as central themes of lean the greatest possible freedom from waste, a focus on value creation from the customer's point of view, internal processes aligned with customer demand, and the integration of employees in the continuous improvement of corporate processes (Gorecki and Pautsch 2013).

Such efforts cannot be reduced exclusively to internal processes. Voigt, for example, points out that the goals of more excellent customer orientation and consistent cost reduction should be achieved both internally and across the company (Voigt 2010).

Pfeiffer also supports this approach with his statement that 'the scope of the Lean Management concept, as we understand it, extends substantially beyond the production area and is relevant for all functions, structures, and processes of a company or an entire value network' (Pfeiffer 1994). Bertagnolli goes one step further in his work *Lean Management*, published in 2018. Here, too, the author reinforces the approach to optimisation and improvement, whereby the methods applied in the production environment can be segmented into the categories of stabilisation, flow, takt, pull, and perfection. However, according to Bertagnolli, the central focus is on more sustainable goals: 'Lean has mainly to do with the strategy and culture.'

From the above examples, it can be seen that the topic of lean is also subject to constant change, which can be attributed to the development of new methods and the changed focus of observation by the applying companies. For example, the beginnings of lean management, which was significantly influenced by the TPS, resulted primarily from the difficult economic situation in Japan after the Second World War. This included the scarcity of resources due to the isolation policy of the USA towards Japan, which required the Japanese economy to make do with limited available resources, but also the peculiarities of the Japanese market, which was small (low production quantities) but demanding (high-quality requirements) and therefore not suitable for mass production along Ford's lines (Parkes 2015).

Parallel to the added methods, the focus of lean also changed with them. For example, works from the 1990s such as *The Synchronous Production System* by Hitoshi Takeda (1995) and *The Secret of Success in Toyota Production* by Shigeo Shingo (1992) deal mainly with very operational methods primarily tailored to production. However, merely copying or adapting these failed. Thus, in the foreword to the second edition of his 2013 book *The Kata of the World Market Leader*, Rother states, 'Concepts alone are not enough'. Therefore, his book focuses on the mechanisms that generate the changed understanding of leadership, management, and culture for permanent innovation that are the basis for a successful application of lean.

It has also become apparent that the lean concept is specific neither to the automotive industry nor to production, but is also being applied in the service sector. Lean is increasingly being used in other sectors, such as lean engineering, lean construction, lean mining, lean medicine, and lean hospitals, as well as in all areas of the company (lean service management, lean selling, lean administration, lean logistics, and lean factory design).

2.3.1 Resource and Flow Efficiency

We can see different approaches to determining this parameter by looking at the flow efficiency. A more general approach is offered by VDI Guideline 4800 Sheet 1 (see Eq. (2.1)). According to this guideline, resource efficiency means the application of an optimal cost–benefit ratio in the use of resources (Verein Deutscher Ingenieure 2016).

$$\text{Resource efficiency} = \frac{\text{Benefit}_{(\text{process, organisation, product, service})}}{\text{Need}_{(\text{use of natural resources})}} \tag{2.1}$$

For the calculation, the data used for the numerator and denominator must subsequently be defined and quantified according to the guideline to calculate it.

Bach et al. refer to another possibility of determining resource efficiency since this one-dimensional view has only a meagre informative value. To evaluate resource efficiency holistically and sustainably, it is necessary to consider the use of raw materials and such complex issues as the environmental impact associated with the extraction and use as well as the availability of the raw material currently and in the future. Bach et al. use the ESSENZ method for this purpose, which includes the dimensions 'physical availability', 'socioeconomic availability', 'social acceptance', and 'environmental impact' in the calculation (Bach et al. 2016).

Resource efficiency in the original sense is of particular importance for the construction industry, as it is one of the most resource-intensive sectors of the economy and therefore has correspondingly enormous potential for savings. In the construction industry, resource efficiency encompasses material and energy efficiency subsectors (Verein Deutscher Ingenieure 2021).

For an organisation, resource efficiency means using the required resources (facilities, personnel, tools, information systems) during the production process so that the maximum utilisation of resources is achieved. The needs of the customer do not have to be in the foreground.

Inefficiencies can arise in pure resource-oriented efficiency:

- Long lead times.
- Many tasks at the same time.
- Many restarts for a job.

Flow efficiency indicates how efficiently a product flows through the organisation based on the view of the customer. It measures how long it takes to identify the customer's needs and to meet them (Modig and Ahlström 2015).

According to Modig and Ahlström, flow efficiency can be calculated by using Eq. (2.2):

$$\text{Flow efficiency} = \frac{\text{Value adding time}}{\text{Total duration}} \tag{2.2}$$

2.3.2 Examples for Resource and Flow Efficiency

2.3.2.1 The Machine and Plant Manufacturer

The following example shows the highly simplified calculation of the flow efficiency for the machining process at a machine and plant manufacturer. The value-adding time is the active machining of the workpiece (spindle running time). The setup time includes all the necessary preparatory work (preparing and setting up the machine, placing and fixing the workpiece, measuring, selecting the appropriate tools and machining program) required to start machining the part.

More precise differentiations of the setup time (as with SMED), as well as possibly occurring disturbances (as with overall equipment effectiveness, or OEE), are neglected in the context of this example (see Table 2.1).

Since the calculated value shows strong correlations with other parameters (e.g. production organisation, production volumes, component geometries, clamping situation, special requirements about work safety, etc.), this value in isolation is only of limited significance. Similar to OEE, internal (before/after an implemented improvement) and external comparisons (benchmarking) are recommended.

Table 2.1 Example: flow efficiency.

Machining process: chipping	
Value-adding time	15 h
Setup time	+2 h
Total duration	= **17 h**
Flow efficiency $\left(\frac{15 \text{ hours}}{17 \text{ hours}}\right)$	= 88%

Suboptimal conditions or isolated solutions arise if each resource is optimised in isolation. In lean management, 'efficiency', therefore, means a fast and smooth process flow with the highest possible degree of utilisation of the resources used, geared to the customer.

2.3.2.2 The Vacation Flight

Resource efficiency is 'normal' when flying with an aircraft, as are the process steps of baggage check-in, security check-in, and boarding. Everything is geared to the resource 'aircraft' and its maximum utilisation. If the process were flow-efficiently geared to customer needs, the customer could complete all the necessary steps at a station without waiting. Still, the airport would have to provide significantly more personnel and infrastructure.

2.3.2.3 The Healthcare System

An individual goes first to the family doctor to diagnose a possible illness, who sends them to the specialist. It takes weeks to get an appointment. When the patient arrives at the specialist, they will receive a referral for radiology. Again, weeks pass, and after the X-ray appointment, weeks pass until the results are discussed with the specialist. As in the first example, this is about resource efficiency. If it were flow efficiency, the patient would go through all the process steps at the first appointment and be diagnosed after three hours, not three months.

2.3.2.4 The Automotive Industry

As shown in Section 2.2, the automotive industry represents the cradle of a resource-optimised and flow-oriented production organisation. Thus, the methods and tools discussed here can still be found there today, even if some of them have been subject to specific adaptations over time. Takt, pull, and flow can be found in every clocked production line in the automotive industry. Whilst at Ford work steps, according to the Taylorist principle, still show a maximum degree of dissection (Taylor 2013 cited in Wirtschaftslexikon24 2018), there has been a move to group individual work steps in a meaningful way and to have them processed by teams working in rotation to counteract the damaging physical and psychological consequences of monotonous activity.

Another example of maintaining cycle time and flow is the immediate rejection of nonfunctional end products with downstream troubleshooting.

2.3.3 Four Important Principles

In 1993, Porsche was in a severe crisis: the company was facing insolvency and hostile takeover. Wendelin Wiedeking, mechanical engineer and Porsche's chief executive officer, achieved a turnaround within six years thanks to his application of the lean management approach (Kippels 1999). The following four principles have been successfully established as a result.

2.3.3.1 Flow Principle

The flow principle maintains continuous flow. Work steps are coordinated so that a harmonised process is created. The focus is on the process and not on the final product. The flow principle is used, for example, when the subsequent work is carried out immediately after the completion of the work of a previous trade.

2.3.3.2 Takt Principle

The takt principle forms the heartbeat of production and indicates the rhythm in which progress is made. Takt in the process means a harmonised and balanced utilisation of production capacity and the avoidance of disturbances in the process.

2.3.3.3 Pull Principle

The pull principle ensures that all resources are included in the process at precisely the right time. This means that the work is not passed on to the subsequent work phase until the next trade is directly in a position to continue the work in production.

2.3.3.4 Zero-defect Principle

Errors found during the process and not after completion mean less waste. Even more important is that the zero-defect principle is used to learn from mistakes and optimise the process. The production process should be continually improved.

These four simple processes:

- flow
- takt
- pull
- zero defects

express a lot of what lean is and are discussed in more detail in the context of 'construction' in subsequent chapters.

2.3.4 Lean Leadership

Like so many other methods and tools in this book, the concept of lean leadership was developed at Toyota, except that the approach is called 'hoshin kanri' there.

A definition of the term in the conventional sense is not initially found by Liker and Convis in their book *The Toyota Way to Lean Leadership*. According to them, the difficulty lies in comprehensively defining or describing good leadership. In contrast, exemplary leadership in lean leadership is easy to recognise, which they illustrate with one positive and one negative example (Liker and Convis 2012).

Nevertheless, specific basic ideas, implementation stages, and fundamental principles are identified and briefly described here before we go on to outline the significance and benefits of lean leadership to the construction industry.

According to Christoph Marquardt, three basic ideas can be identified in lean leadership (Marquardt 2018):

- The continuous improvement of all processes on the one hand and the human being as such on the other hand are at the centre of these considerations.
- To do justice to this human-centeredness, managers are prominent in implementing lean leadership. They are trained to guide their employees in their area of responsibility to further development. The goal of further development is, on the one hand, to increase the problem-solving competence of employees and, on the other hand, to enable them to improve all corporate processes continually.

2.3.4.1 Excursion: Kata

This is done, among other things, through coaching kata, which will be briefly discussed in this context. The Japanese word *kata* initially describes an exercise that martial artists constantly repeat. Analogously, Rother describes the approaches of the improvement and coaching kata in his work *The Kata of the World Market Leader* (2013). The coaching kata tool is necessary to systematically involve employees in the continuous improvement process and ensure they apply improvements or the improvement kata in their daily work. The employees to be developed (mentees) are supported by coaches, coordinators, teachers, or trainers (mentors) who help them gradually approach the desired target state.

In addition to the further development of personal problem-solving skills, this is also an overriding objective in general. The achievement of corporate goals is transformed into a process perceived as natural.

Companies that implement lean leadership take a systematic approach to identify and develop their leaders. They are guided by the leadership development model, according to Liker and Convis (2012), that includes the following stages, as can be seen in Figure 2.9.

First Stage This stage is about developing oneself as a leader. It is essential to reflect and develop one's own leadership performance. The principle of the 'reflective practitioner' from Donald Schön (2008), a former MIT professor, should also be mentioned. The concept considers self-reflection competence an essential quality for coping with complex situations.

Second Stage This stage is about coaching and developing others. It is essential to create other people as leaders to acquire the competence to reflect on their own performance to establish themselves.

Figure 2.9 Toyota leadership development model according to Liker and Convis (2012).

Third Stage Stage one and two are about the individual development of the people and the leader. The third stage focuses on the institution and aligning teams, departments, and divisions in the direction of daily kaizen to systematically ensure and maintain improvement.

This stage is about creating a vision and coordinating the goals. The whole organisation must be aligned to hoshin kanri ('management by policy' or 'policy deployment'), which means that there is no silo thinking in the divisions but maximum transparency of the production processes flow-oriented across all departmental boundaries. The (divisional) goals must be coordinated and focused on achieving the top corporate goals with a corresponding resource allocation.

Despite this structured approach to implementation, the essence of lean leadership is still hard to define. Nevertheless, certain principles of a promising lean leadership approach can be identified. Dombrowski and Mielke (2013) were able to identify five basic principles in their research of the literature on this topic. These include:

- improvement culture
- self-development
- qualification
- gemba (shopfloor management)
- hoshin kanri.

Finally, it is essential to be able to explain to the interested reader as well as to stakeholders in the construction industry the advantages of the implementation of lean leadership and why it is necessary.

Similar developments and problems which influence leadership and leadership quality observed in the construction industry can also be seen in numerous other industries.

These include, in particular, the shortage of skilled workers in general and recruitment problems in many companies. These result mainly from the working conditions in construction, which include the physically demanding nature of the job and the direct impact of weather conditions, and a somewhat rustic culture of discussion and management. Another factor is the often inadequate training and introduction of junior managers to their tasks, which involve mainly troubleshooting rather than active management.

Under such conditions, the assumption of management tasks, which often include employees from other companies or service providers but with very limited time, is often a challenge that is difficult to master.

Under such conditions, it is not possible to realise planned management, an appreciative leadership culture, or proactive process improvement.

It is precisely these problems that lean leadership addresses with its changed understanding of leadership.

While the detailed design of a possible lean leadership implementation project is not be discussed in the following, the potential changes of a successful implementation of lean leadership are.

For example, Marquardt (2018) cites several positive changes as a result of implementing lean leadership at the worksite, such as significantly reducing the need for written and verbal communication with business partners in the context of clarifications. (One of the authors can confirm the procedure and the described result; see the case study in Chapter 3, Section 3.5.2).

It was also found that problems and deviations were identified more quickly, resulting in a significant reduction in the total number of anomalies. This was also because identified issues could generally be addressed more quickly. This had the positive effect that problems were not forgotten and dragged out, which could have had a negative impact on further construction progress.

Marquardt also mentions improvements in the personal area. For example, construction managers would have felt less emotional stress in some cases due to the more coordinated daily routine and the lower number of misunderstandings. An additional positive factor would be the faster-growing together of the construction site team into a real team.

Despite all the positive aspects mentioned, it should not be forgotten that such a far-reaching change in the management and organisation of a construction site also brings with it difficulties when adapting to new processes.

Just one example can be cited here. The new role as moderator and mentor at the daily cycle meeting takes some getting used to for the responsible site manager (Marquardt 2018).

In conclusion, it can be said that lean leadership requires a sustainable change in the way employees and managers work together. Or as Dombrowski and Mielke (2013) state: 'A lean leader needs to be a role model for their employees.'

2.4   Lean Construction and Tools

Owing to the achievements of lean management, especially outside Japan, it was only a matter of time before the adaptation of the management approach, which is rooted in the TPS, would take place in other industries. Lean construction is now the corresponding transfer to the construction industry, although the primary goals of the approach have not changed (Fiedler 2018).

The reasons for adapting the lean approach to the construction industry include external factors comparable to those for the original implementation of the automotive industry.

These concern the economic framework conditions: high demand (and thus increasingly scarce resources and rising prices), a persistently low price level about contract awards, and increasing competition and further political/normative factors.

Therefore, the demand for renewable energies, energy and resource conservation, as well as new quality standards have been around in the construction industry for quite some time (Fiedler 2018). In addition to the undeniable positive effects that these developments bring, there is also a growing need to raise the efficiency of the industries to a new level using new methods, i.e. lean construction.

Despite the changing environment, the basic demands and goals of lean in the construction sector are the same as in the original formulation of the approach:

- Maximise customer value.
- Continuous improvement of processes.
- Elimination of waste.
- Application of flow and pull principle.

Or, as the Lean Construction Institute (LCI) states, 'Lean Construction extends from the objectives of a Lean Production System – maximise value and minimise waste – to specific techniques, and applies them in a new project delivery process' (Lean Construction Institute 2015).

In the context of construction projects, waste can be found in criteria that arise from a wide variety of stakeholders in a project, such as:

- Insufficient communication and collaborative work within or between different teams. This can result from a lack of operational management on-site and from the fact that international teams are often deployed, especially in large-scale and megaprojects, where direct communication is often impossible without problems due to language barriers.

- A low price level in budgeting results from increased competition, with raw material and employment costs rising sharply at the same time. This can trigger a whole series of negative effects that further increase the burden on or even overspending of the available budget.
- Legal proceedings against projects, which tie up a whole bundle of organisational and financial resources and those directly involved in the project or can even postpone the project indefinitely.

Figure 2.10 shows a small example. In step 01, the activities of the trades are planned successively. The actual duration corresponds to the target duration. During the project (step 02), the time of the former trades is already extended, so shorter durations are set for the trades to keep to the planned schedule. Here the actual time is already longer than the target duration. The mutual interference of the trades causes a kind of freefall in the last step. To survive the project with limited damage, all other activities are parallelised. The result is a loss of quality

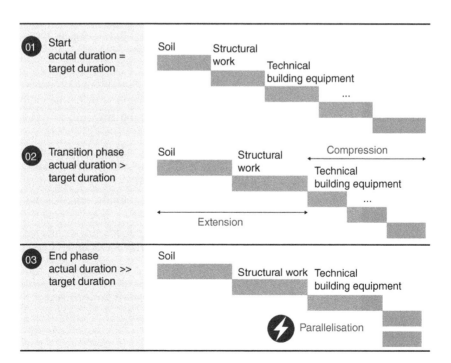

Figure 2.10 Postponement of the time sequences of the trades.

and overruns in time and most probably also in budget. In summary, the following problems can result from this forced parallelisation:

- Mutual obstruction of trades.
- Increasing need for control by site management/polishers.
- Quality problems in the product.
- Danger of exceeding the construction time.
- Explosion of costs.
- The costs of rectifying defects and the energy required to do so rise sharply.

The example clarifies that one of the core problems of the construction industry is the insufficient short cyclical adherence to deadlines in the course of the project. To overcome this challenge, it is necessary to switch from a classical approach, which is directed top-down at the construction site, to a bottom-up or process-oriented approach. A holistic value-added flow is generated, and short cyclical deadline compliance is possible. In other words, the construction site must be the centre of attention.

All secondary activities of construction companies – such as acquisitions, logistics, or control –must be oriented towards a holistic value-added flow so that the final place of value creation, namely the construction site, is placed at the centre of the process. This is also structurally required with the Viable System Model (see Chapter 1). The saying 'The purpose of the system is what it does' reminds us that, ideally, the construction site should be about how to build efficiently and effectively – and nothing else.

Process-oriented thinking and the principles of the lean philosophy must be tailored to the specific needs of the construction industry. Simply copying the tools does not ensure successful implementation. Furthermore, the principles must be adapted. For example, in stationary industries, such as the automotive industry, workstations are locally bound, and the product moves through the workstations. In contrast, the development of a building is locally attached to a construction project. The trades move through the building. If lean thinking is successfully adapted and implemented, it is possible to increase value-adding activities and minimise waste.

Interest in lean construction is growing continually as an increasing number of owners, contractors, and consultants focus on lean construction. Lean strategies of companies as well as of international and national lean construction institutes reflect the recognition that lean thinking has arrived in the industry. Meanwhile, a broad portfolio of lean methods and tools for the construction industry exists.

2.4.1 Last Planner System

In 1997, two American civil engineers, Glenn Ballard and Gregory Howell from the University of Berkely, founded the Lean Construction Institute. They thus laid the foundation for lean process planning in the construction industry. Since then, the last planner system (LPS) and takt planning and control have established themselves as essential practice methods.

The LPS is a tool developed by Ballard and Howell to stabilise planning and control processes during production (Ballard 2000). Obstacles are to be overcome with the help of foresighted and short-cycle planning. It is a holistic and collaborative approach to increase productivity.

According to the German Lean Construction Institute (2019), compared to conventional construction management, the LPS delivers the following added value:

- The satisfaction of all project participants, both internal and external, is increased.
- Projects become more predictable and stable.
- Workflows and processes are organised more reliably and 'fluently'.
- Milestones are reliably met, and production times are reduced.
- The final product is achieved with greater safety and quality.
- Project management is achieved through the integration of the project participants.
- The outdated approach of pure subcontractor control is transformed into self-responsible target compliance.
- Stress and 'fire-fighting actions' are reduced.
- Communication and transparency during the execution are improved.
- The added value of projects as a whole is increased.
- LPS supports the successful implementation of further lean management measures.

The last planner role is decisive for the LPS. The person supervises the planning or execution process as the last instance. In construction processes, the foreman is often the last planner. As a local manager, he is directly involved in the construction process and is not as contract-focused as the site managers. In the case of LPS in design, the inspectors, test engineers, or subsequent planning trades are involved in the design process at a late stage. According to LPS, these must be included as early as possible.

The LPS works according to the pull principle mentioned above (see Section 2.3.2). This means that all subprocesses are controlled according to requirements to guarantee a smooth transition in terms of materials, deadlines, and quantities. The aim is that subsequent instances can begin directly with their work packages.

The opposite of the pull principle is the push principle, which is in reality implemented on most construction projects. Regardless of whether the respective work has been completed or not, the next activity is 'pushed in'. Each trade only takes care of its interests; the overall optimum is unimportant. The view for the whole and the collaborative coordination of the project participants are missing.

With the pull principle, all activities are oriented towards the overall process and viewed in a continuous flow. This means that all activities are planned JIT, and the collaborative approach creates a stable flow. The existing buffers of all participants are disclosed and can be used by the overall system.

The LPS is divided into five phases; see a short example in Figure 2.11.

2.4.1.1 Milestone Planning
A classic schedule serves as the basis for process planning: milestones or quality gates are defined from this schedule for the entire project duration by the customer/client or general contractor.

2.4.1.2 Collaborative Programming
Together with all project participants (client, general contractor, subcontractor, site manager, planner, etc.), all work packages and the production sequence are determined. Interdependencies between the different trades are highlighted, and the buffers of the individual companies are made transparent and used for the overall system. This optimises processes, and the comprehensive system becomes more stable.

2.4.1.3 Making Ready
Tasks are defined for the coming weeks. Prerequisites are determined, and adjustments are made if necessary. In addition, work packages are worked out in more detail, alternative tasks are defined, and functions that still need to be completed are listed in a to-do list. The aim is to prepare the work packages so that they can be carried out realistically.

2.4.1.4 Production Planning
Here the work for the next period (week, day, shift) is agreed upon together. The work packages have a high level of detail, and all prerequisites are fulfilled to carry them out. The detailed planning takes into account the knowledge of the performance already achieved. In daily meetings, it is checked whether the work can be carried out as planned. If changes are required, adjustments are made.

2.4.1.5 Production Management and Learning
Next to managing the production, the processes should be improved. The aim is to learn from mistakes and to improve processes.

The LPS is used to measure compliance with commitments and the reliability of stakeholders. The planned percentage complete (PPC) is the associated measurement criterion. Reasons for the noncompliance are often:

- Lack of preliminary work.
- Lack of communication.

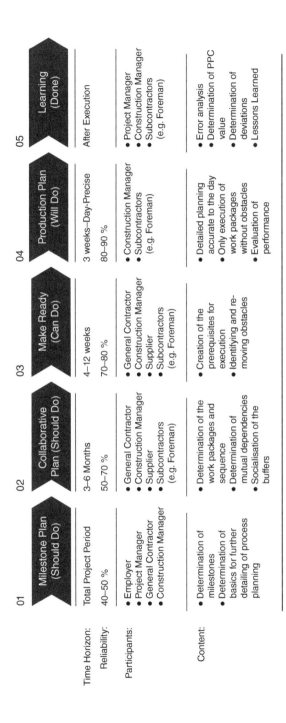

Figure 2.11 The last planner system.

- Lack of personnel or equipment capacity.
- Lack of quality.
- Missing permissions.
- No responsible person in the team.

2.4.2 Takt Planning and Control

The takt is a basic temporal structure in music. The individual beats structure the music; this structural specification is also essential in manufacturing products. Takt production is a special form of flow production and was used by Henry Ford at the beginning of the twentieth century. Also, Toyota recognised the importance of takt production and implemented it into the TPS (Haghsheno et al. 2016).

The takt forms the 'heartbeat' of production and indicates the speed at which the work steps are carried out. The application is particularly suitable for frequently manufacturing recurring or similar operations or sequence sections (e.g. carriageways, tunnels, high-rise buildings, bridges).

In general, the takt time can be defined and calculated with Eq. (2.3) (Brunner 2011; Reitz 2009):

$$\text{Cycle Time} = \frac{\text{Daily production time}}{\text{Daily production volume (demand)}} \tag{2.3}$$

A harmonised construction process is created based on JIT, the four lean principles (flow, takt, pull, zero defects), and further Toyota principles. The work for one cycle is planned so that a smooth process, constant activity with minimum idle times, and a guarantee of executability in the available time are created.

According to the German Lean Construction Institute (2019), compared to conventional construction management, takt planning and control delivers the following added value:

- An open communication culture is created within the project team.
- Construction processes become transparent, measurable, and predictable.
- The focus is shifted to the creation of value.
- Problems and obstacles are identified before execution and can be solved and eliminated in time.
- All those involved in the construction process are involved in the process at an early stage with their experience.
- Construction work is harmonised, planned, and carried out in a precise rhythm or cycle.
- The effect of continuous improvement occurs. The goal is self-learning and continuously improving the organisation.
- Productivity is increased in all project phases.

- Projects can be realised on time, within budget, without defects, and with stable processes.
- The satisfaction and stability are increased for all project participants.

The takt plan is used as a basis for tendering works. This is checked by the potential contractors and adjusted if necessary. This lays the foundation for a standardised process that defines the framework conditions for all parties involved.

These general explanations of takt planning and takt control in practice will now be illustrated in more detail with general examples from the construction industry.

2.4.2.1 Takt Planning

Different approaches have been developed for creating a takt plan. For instance, Dlouhy et al. (2018) at the Karlsruhe Institute of Technology and Frandson et al. (2013) have published different takt approaches.

Overall, takt planning includes the following steps (see Figure 2.12).

Zoning of the Construction Project The building project is divided into functional areas. Zones are prioritised according to the customer's needs. All zones are subdivided into the smallest repeatable element or the smallest common multiple. The smallest common multiple can be, for example, a housing unit, a hotel room, or a section of a road. Its formation enables the standardised duplication of takt plans to other areas.

Definition of Workload All necessary tasks and trades for the smallest common multiple are determined. In addition to the work content, the effort is also calculated in terms of resources.

Developing of Trade Sequence This is followed by the definition of a logical sequence of trades. It is determined which work packages must be processed

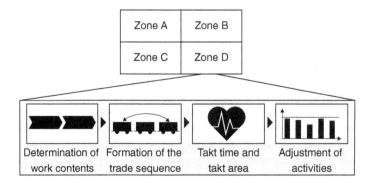

Figure 2.12 Takt planning.

consecutively and which can be processed in parallel. Here it must be considered which trades require which preparatory work.

Definition of Takt Time and Takt Area The required takt times and areas are determined considering the completion date and other milestones. The usual duration for a takt for construction projects is one week.

Adjusting Activities The finalisation of the trade sequence follows this. The various activities are adapted so that all trades can maintain the takt time. A bottleneck station determines the takt time for the entire process chain. To smooth the activities in the takt, tasks can be bundled, the workforce increased, or the additional use of machines can be planned. The aim is to ensure a balanced workflow.

Determining the Takt Strategy and Production Schedule Depending on the type of building, the number of floors, and the architecture, a strategy has to be defined to determine which direction the trade train should run along. In addition, the number of possible trades to be used is discussed in the cycle strategy. The strategy is developed about the final deadline and creates a production schedule.

Repetition of the Steps and Creation of the Takt Plan The last step repeats the previous steps by determining the takt time and the takt areas for all functional zones. The results are displayed in the takt plan. If more significant deviations occur, takt stops, or empty takts are taken into account.

2.4.2.2 Takt Control

During the construction process, the construction site is coordinated through the use of takt control.

Takt control meetings are held daily to reflect the performance of the previous day and to discuss upcoming work packages. In the meetings, topics for the construction site can be dealt with in a short cycle.

The control board makes it possible to share important information with those involved in the project. This leads to transparency. By tracking key figures and checklists, deviations can be detected quickly and countermeasures can be taken. See also Chapter 3's Section 3.4 on performance measurement. A control panel can look like the one shown in Figure 2.13.

2.4.3 Last Planner System and Takt Planning and Control

Two lean construction approaches for production planning have become predominantly established in the construction industry. On the one hand, users of the LPS and, on the other hand, users of takt planning and control (see Figure 2.14).

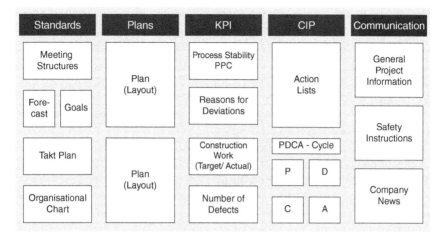

Figure 2.13 Takt control board, graphic aligned on Demir/Theis, Drees & Sommer (Demir and Theis 2018).

Figure 2.14 Last planner system and takt planning and control.

In their approach, both systems strive for a holistically optimised value-added process from the customer's point of view. In addition, short-cycle feedback loops are an essential instrument for continuous improvement. The entire construction process is described in great detail and should occur smoothly and in one flow.

The LPS is distinguished by a particularly collaborative approach, in which the production plan is developed step by step together with all participants. This allows the expertise of all experts to be incorporated and more realistic work values to be determined. The needs of the project participants can be taken into account during scheduling. In addition, the step-by-step further formulation of the production plan allows additional work and deviations to be dealt with in good time, resulting in a higher degree of flexibility. The participants make promises, and a particular group dynamic keeps the promises.

Takt planning and control offer a very detailed preliminary planning and process planning in an early performance phase. Therefore, the process execution is fixed from the beginning and includes a very exact determination of all effort values and external factors before awarding the contract. Owing to the high degree of standardisation, takt planning and control can quickly implement specific work steps and enable a continuous workflow. This leads to stability and security for all parties involved. The challenges of takt planning and control are planning in great detail during the early project phases with little information and implementing this consistently and with methodical discipline.

The origin of the TPS has JIT (aligning with customer's needs) and jidoka (transparency during the production process) as its goals.

Takt planning and control adapts the systems and methods of automobile manufacturers to the construction industry. The LPS interprets the lean principles in an agile way and aims primarily at cooperative and collaborative process planning. In summary, the LPS is more of a bottom-up and takt planning and control more of a top-down approach. Depending on the framework conditions and the task, either one or the other, or a combination of both, is suitable for the application.

The practice has shown that many companies have designed their production systems. An example of a hybrid system is lean construction management practised by the project management company Drees & Sommer. One method was rigidly used, but a hybrid system consisting of takt and LPS elements was subsequently designed that best suited the organisation. This depends strongly on the company's performance or production method and is determined according to demand. Organisations that can achieve a high degree of prefabrication benefit from detailed production planning at early stages, ensuring the production process's stability. Because of the many uncertainties, organisations that need agility will achieve higher productivity with the LPS.

Digitalisation is also making rapid progress in using these approaches, as shown by applications such as LCM Digital, Yolean, Bosch Refine My Site, Refine VVC, or Koppla and Takt.ing. The intelligent use of data, automation, integration into the BIM environment, and the use of partnership-based contract models will further optimise and sustainably improve planning and construction processes.

2.4.4 Lean Construction Case Study

The following is a simple example of a construction project applying takt planning and control for interior finishing.

Project boundary conditions: A student residence with 200 apartments in four buildings. Gross floor area is 6000 m^2. Planning and structural work on the dormitory has already been completed. The interior construction of the building project with the help of takt planning and control is to be executed.

Project scope: Approx. €20 million.

Project duration interior finishing: Four months.

Main trades involved:

- drywall construction
- electrical trade
- fire protection trade
- heating, ventilation, air conditions (HVAC) trade
- painting trade
- window trade
- smaller trades.

2.4.4.1 Takt Planning

Zoning of the Construction Project   The first analysis of the construction project shows that the building has a high number of repeatable units. The structure of the four buildings is almost identical so that the production planning for one building can be transferred to the other three buildings. By looking at the side view (Figure 2.15), it will be clear that all floors except the ground floor show repeatable elements. A closer look at the floor plan (Figure 2.16) reveals that mainly single rooms with the exact dimensions are produced. The staircase and a flat share room are the only special constructions per floor. A single room can be used as the smallest common multiple.

Definition of Workload   After the zoning is completed, the workloads of the trades can be determined. There is a database with effort values for all services that they

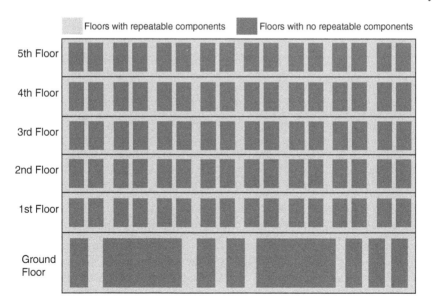

Figure 2.15   Side view student housing.

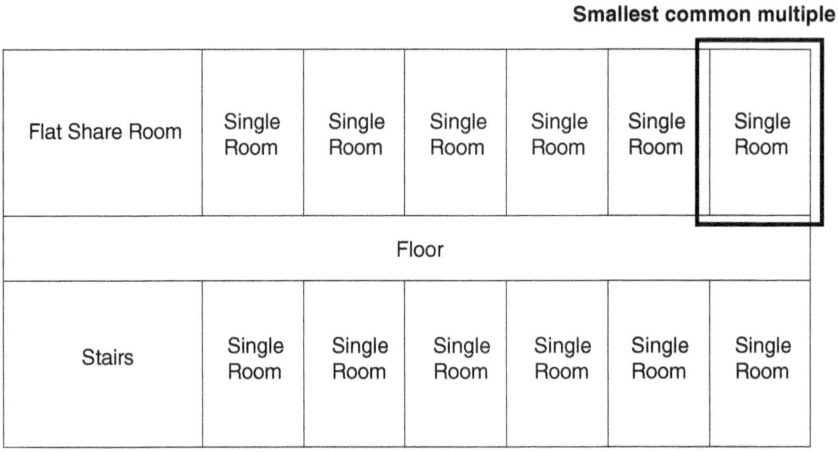

Smallest common multiple

Flat Share Room	Single Room	Single Room	Single Room	Single Room	Single Room	Single Room
Floor						
Stairs	Single Room	Single Room	Single Room	Single Room	Single Room	Single Room

Figure 2.16 Floor plan view.

Trade painting

Activity	Quantity / Mass	Effort value (h/unit)	Man hours (h)	Staff required	Duration (h)	Total duration (d)
Create color	30 l	0.1	3.0	1	3.0	
Paint walls	80 sqm	0.2	4.0	2	2.0	0.81 days
Paint sockets	15 units	0.1	1.5	1	1.5	

Figure 2.17 Example determining work content of trade painting.

have to fulfil for the trades. In addition, the workload values are confirmed by the foremen of the trades, thus ensuring that the figures are not obsolete. An example of the procedure is shown in Figure 2.17. The result is the estimated total time (in days – by eight hours/workday) for each trade.

Developing a Trade Sequence The next step is to define the sequence of trades for the smallest common multiple. Together with the specialist trades, the preparatory work required for which business is decided. By mutual agreement, a sequence is determined so that each transaction can carry out its activities without additional waiting times. The chosen sequence is shown in Figure 2.18.

Figure 2.18 Determining trade sequence.

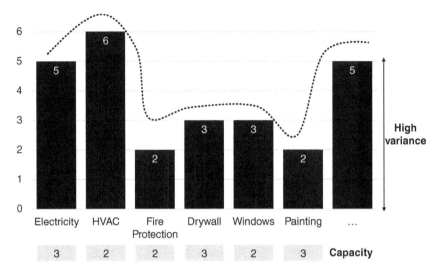

Figure 2.19 Mapping trade sequence with workload of trades.

In the next step, a work distribution diagram is generated. This takes into account the effort values and capacities from the previous step for each trade. The chart shows a high variance between the cycle time of the different trades. This means that each trade delivers its output at a different rate per smallest common multiple.

More specifically, it is found that some of the trades require between two and three days. Another two trades require five days. The HVAC trade turns out to be the trade with the highest daily duration of six days (see Figure 2.19).

Definition of Takt Time and Takt Area and 5. Adjusting Activities Based on the work distribution diagram and experience values, a cycle of five days is determined. This has the advantage that, on the one hand, rework can be postponed to the weekend if necessary and, on the other hand, that the materials can be dried over the weekend.

In addition, the trades are consulted, together with the HVAC trade, to see how the duration can be shortened. One of the foremen comes up with the idea that the duration can be reduced from 6 to 4.5 days by using a new machine. This means

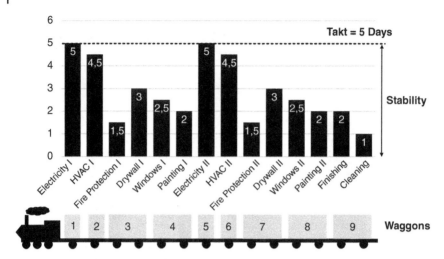

Figure 2.20 Work distribution diagram including takt time.

that the trade is ultimately within the takt time. The trades that require considerably less than five days for their work are combined into one trade. Thus, for example, the fire protection and drywall trades are put into one wagon, which optimally uses the weekly capacity utilisation. In total, nine harmonised 'takt' wagons are determined (see Figure 2.20).

Based on the takt, the takt areas are defined. The takt ranges consist of the smallest common multiples. It is decided that two single rooms will form one takt area for this building project, based on the floor plan. In addition, the stairs and flat share room areas are each defined as one takt area (see Figure 2.21).

This results in eight takt areas per floor (excluding the ground floor).

Determining Takt Strategy and Production Schedule The next step is to develop a construction strategy for production planning. It is determined that the interior work will begin on the ground floor and then each floor in turn will be completed for the construction project. Owing to the available resources of the subcontractors, two trains can be used. This makes it possible to start simultaneously in building 1 and 3. A building direction is also determined for floors 1–5. All other single rooms will be completed beginning with the flat share room. The last step is the staircase (see Figure 2.22).

Repetition of the Steps and Creation of a Takt Plan The work steps described above are carried out repeatedly to generate the takt plan (see Figure 2.23) or production schedule.

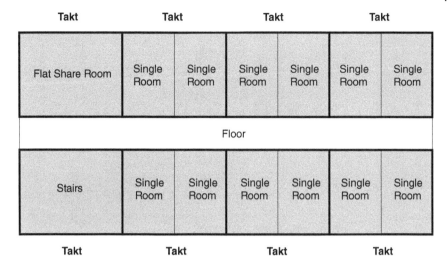

Figure 2.21 Takt areas.

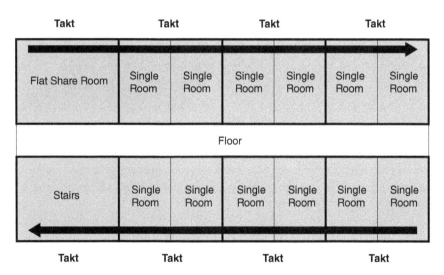

Figure 2.22 Takt strategy.

2.4.4.2 Takt Control

A takt control board and takt control meetings are introduced during the construction phase based on the takt planning. The control board contains all essential information required for a stable production process. General information such as meetings times, project participants and responsible persons, and telephone

Building	Floor	Train	01	02	03	04	05	06	07	08	09	10	11	12	13	14	15	16	17	18
	1		W1	W2	W3	W4	W5	W6	W7	W8	W9									
	2			W1	W2	W3	W4	W5	W6	W7	W8	W9								
1	3	1			W1	W2	W3	W4	W5	W6	W7	W8	W9							
	4					W1	W2	W3	W4	W5	W6	W7	W8	W9						
	5						W1	W2	W3	W4	W5	W6	W7	W8	W9					
	1							W1	W2	W3	W4	W5	W6	W7	W8	W9				
	2								W1	W2	W3	W4	W5	W6	W7	W8	W9			
2	3	1								W1	W2	W3	W4	W5	W6	W7	W8	W9		
	4									W1	W2	W3	W4	W5	W6	W7	W8	W9		
	5										W1	W2	W3	W4	W5	W6	W7	W8	W9	
	1		W1	W2	W3	W4	W5	W6	W7	W8	W9									
	2			W1	W2	W3	W4	W5	W6	W7	W8	W9								
3	3	2			W1	W2	W3	W4	W5	W6	W7	W8	W9							
	4					W1	W2	W3	W4	W5	W6	W7	W8	W9						
	5						W1	W2	W3	W4	W5	W6	W7	W8	W9					
	1							W1	W2	W3	W4	W5	W6	W7	W8	W9				
	2								W1	W2	W3	W4	W5	W6	W7	W8	W9			
4	3	2								W1	W2	W3	W4	W5	W6	W7	W8	W9		
	4									W1	W2	W3	W4	W5	W6	W7	W8	W9		
	5										W1	W2	W3	W4	W5	W6	W7	W8	W9	

Calendar weeks

Figure 2.23 Takt plan.

numbers are listed on the board. The board also contains the construction process and an excerpt of the takt plan. Both documents need to be readable and understandable for all parties involved.

A further component of the board is listing all relevant deadlines and milestones. This is marked if a service is not provided on time by the trades. In addition, joint consideration is given to how the performance can be maintained with countermeasures.

Another essential part of the takt board is the inspection of quality and cleanliness on the construction site. These factors are tracked with key performance indicators (KPIs).

The second decisive driver of takt control are the takt control meetings. These take place daily in this project. In this context, coordination with the trades is carried out to ensure that they can carry out their tasks smoothly. Furthermore, the performance already achieved is reflected upon. In addition to the daily meetings, a joint meeting on the control board is scheduled every two weeks to ensure the process's stability and to optimise the existing process further.

2.5 Tools, Tools, Tools

Applying lean construction to projects can be supported by using several tools which have their origins from different industries or were designed especially for the construction industry. These tools are used mainly for:

- project and process analysis
- organisation and control
- continuous improvement and learning.

Such tools are especially helpful in bringing the new philosophy and lean thinking to the fore. Furthermore, it is essential to establish such devices in the project to support the introduction of lean processes and make them as simple as possible so that lean users can directly recognise their added value. In the best-case scenario, lean tools are standardised and created as templates, as shown in Section 2.5.4 with the A3 method and report. The templates must be easy and unambiguous to use (according to poka-yoke, errors are to be avoided preventatively). These methods and tools are listed here but, owing to the countless number of lean tools, not exhaustively.

The authors' experience has shown that this often leads to a discussion of whether lean tools should be designed in analogue or digital form. As a rule, it is advisable to introduce analogue tools first and improve them constantly. During this phase, it is essential that the tools can also be used on the construction site. For example, it is a good idea to laminate the documents to be filled in more often with a whiteboard pen. Once the papers are conceptually mature and have been validated by the users, the next step is to consider how the tools can be designed digitally. The prerequisite for this is the existing digital infrastructure.

Overall, there must be a continuous flow in terms of flow efficiency, both analogue and digital. This ensures that the tools do not create additional work or disturbances in the process but rather support the workflow and the application of the lean methodology.

Compared to the construction industry, the combination of lean with Industry 4.0 is already well advanced, especially in the mechanical engineering sector, where the focus is on improvements in production and productivity in general and the interfaces to sales, engineering, purchasing, and logistics.

In the VDMA guide 'Industry 4.0 meets Lean' (VDMA Forum Industrie 4.0 & PTW Institut für Produktionsmanagement, Technologie und Werkzeugmaschinen 2018), the VDMA Industry 4.0 Forum and the PTW Institute for Production

Management, Technology and Machine Tools examines the basis of this topic and also presents best practices from the industry, which are also likely to be groundbreaking for other sectors.

As examples of these best practices, the VDMA lists (2018):

- Direct digital transfer of job configuration data to the associated CNC program to reduce transmission errors. Parallel comparison of the configuration with existing geometries to reduce product variance.
- Visualisation of communication requirements by coloured light signals and notification on various electronic devices used in production. Overview of all workstations and the assigned 'traffic light phases' via a central dashboard.
- Connection of maintenance and digitalisation using an app connected to the central maintenance system. The advantages of this are immediate access to the required documents, creation of photos or documentation with little effort on site, information about stock levels of the necessary spare parts in real-time, time savings, and increased consistency of the data recorded in maintenance.
- Digitalisation of all production processes and visual and store floor management tools. Here, too, the focus is on reducing the time required and making all relevant information permanently available to all interested parties in real time. In addition, consistent digitisation lowers the inhibition threshold for using the tools, promoting a lean culture.

2.5.1 First-run Study

A first-run study is a joint simulation or execution of a critical process. It can be used to optimise any activity. A first-run study is especially recommended for repeatable elements (for example in cycle planning) to achieve significant savings potentials (Herrera and Munoz 2015).

The procedure is oriented on the PDCA cycle. However, it involves the phases plan, do, study, adjust.

2.5.1.1 Phase Plan

In the beginning, a flow chart of the selected process is designed by all parties involved (construction management and craftsmen). The diagram serves as a kind of script. Afterwards, process improvements are discussed based on the flowchart and then incorporated. The new process is checked to record costs, deadlines, quality, and countermeasures. In the last step of this phase, parameters and aspects to be observed are determined, which should be explicitly considered in the first run.

2.5.1.2 Phase Do

After planning, implementation takes place. If possible, it is beneficial to record the first run as a video so that the analysis can be done even easier afterwards. Furthermore, it is essential to record all actual results for the next phase.

2.5.1.3 Phase Study

After the results have been compiled, the results should be discussed with all parties involved. For the group discussion, it is beneficial if the results are prepared visually. For instance, the first run can be divided into wasteful and value-adding activities. During this phase, it is essential to clarify for the participants that any mistakes should be considered treasures. Finger-pointing is out of place here and in the lean philosophy generally.

2.5.1.4 Phase Adjust

Once the problems and causes have been identified, the next step is to work together to improve the process. The suggestions for improvement are discussed, and measures for optimisation are developed and initiated.

If the approach in the context of the first-run study is transferred to the product level, parallels to the minimum viable product (MVP) arise. This method of agile product development originating from Silicon Valley gives the applying company (primarily start-ups) a competitive advantage due to the faster implementation of innovations or new requirements. The method is characterised by an iterative cycle consisting of developing an MVP, measuring it against real users, learning based on customer feedback, and repeating (adapting) the MVP (Depiereux 2021).

2.5.2 Waste Walks

Waste walks, or muda walks, are a simple method for analysing production and construction sites to identify potential for improvement or waste and address the need for action (see 'Waste' in Section 2.2.1 and 'Go to gemba' in Section 2.2.2).

According to the guideline VDI 2553 in Verein Deutscher Ingenieure (2019), before the waste walk begins there are preparations to be made. First of all, it is advisable to create a general overview. Involved companies, logistic concepts, and further information are helpful to perform a waste analysis.

Then the scope of the analysis has to be determined. For example, the entire construction site or only selected sections of a construction site can be analysed. The inspection can be done by one person or in a small group. The categories for waste are based on the classic types of waste. In the best-case scenario, the inspection is repeated at different working times to obtain a representative observation.

Particular attention should be paid to the following points (Liker and Meier 2006):

- How was the deviation identified?
- On what basis could the person making the determination identify this as a deviation?
  - By deviation from a defined standard.
  - By experience.
  - By intuition.

- Could the problem be solved by the person who identified it, or did they seek further help?
- Was there already a defined solution for the problem when it occurred?

The identified wastes are collected in a recording list and supplemented with images. If possible, quantitative information is added to evaluate it in diagrams.

Subsequently, the identified waste elements from the inspection are sorted according to the estimated potential, and measures for improvement are defined.

After defining countermeasures, it is essential to follow the results and derive further adjustments if necessary.

Waste walks can be assigned to S3* of the VSM. Here, for example, the site manager or foreman walks through a specific construction site area to identify waste in the current work steps.

2.5.2.1 5 Why and 6W Questioning Technique

The 5W method stands for 'five times why' and is a problem-solving technique, which is only part of the 6W questioning technique. In this questioning technique, the six questions 'What?', 'Who?', 'Where?', 'When?', 'Why?', and 'How?' are asked six times in a row. The goal of this extensive questioning, which is a unique Japanese characteristic, is the systematic search for the root of a problem that has occurred (Brunner 2011).

The fivefold question of 'why' takes a prominent position here. For this reason, the 5W method is often found in isolation in the literature.

According to Bertagnolli, 5W is symbolises the intensive questioning of a problem or issue until its root is revealed, since this may not be the case even after five rounds of questioning (Bertagnolli 2018).

Additionally, 5W only analyses one root cause. Usually, there are several problems on a construction site, so it is advisable not to limit oneself to only one question word.

If 5W is preferred, it is necessary to identify all existing problems. Furthermore, the results are not repeatable. Different people using 5W would develop other causes for the same problem.

Below is a simple example of 5W.

Problem   The vehicle will not start.

1st Why: The battery is dead.
2nd Why: The alternator is not functioning.
3rd Why: The alternator belt has broken.
4th Why: The alternator belt was well beyond its useful service life and not replaced.
5th Why: The vehicle was not maintained according to the recommended service schedule (root cause).

Another example for a construction project is described below:

Problem   Settlement cracks are present in concrete.

1st Why: Why are settlement cracks in concrete?
Because the company carrying out the work did not make the concrete properly?
2nd Why: Why was the concrete not made correctly?
Because the technical specifications were insufficient, the company worked to the best of its knowledge.
3rd Why: Why was the executing company not in possession of the required documents?
Because the site manager assumed that the documents handed over to him were complete, but they were not.
4th Why: Why did the site manager pass on the incomplete documents to the company carrying out the work?
Because he did not know better and was not trained in this field.
5th Why: Why did the site manager lack this knowledge, and why was he not trained?

Because the company does not have an adequate training system for its employees.

2.5.3   Ishikawa Diagram

The Ishikawa diagram (see Figure 2.24), also known as the fishbone diagram or root cause diagram, is a graphical representation to search for the causes of production errors effectively. All reasons should be identified and their dependencies

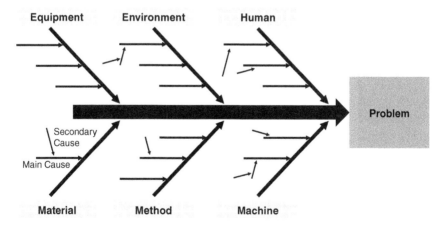

Figure 2.24   Ishikawa diagram.

displayed. In other words, the Ishikawa diagram depicts the cause and its effects on a problem. The chart was developed by Kaoru Ishikawa in the 1940s and is a well-known tool from the TPS (Heller and Prasse 2018).

The basic structure is a horizontal arrow pointing to the right, with the problem or target at the top. This arrow is supplemented by oblique arrows of the main influencing factors that lead to a specific effect. Main influencing factors must be determined depending on the situation. Usually, the influencing factors man, machine, material, method, environment, and equipment (5M) are used.

In the next step, primary and secondary causes are worked out. By using creativity techniques, the potential causes are investigated. The leading causes are placed as small arrows on the main influencing factors. If secondary reasons exist, they are placed on the leading causes. This can be continued arbitrarily so that a ramification arises (Bertagnolli 2018).

Finally, the diagram must be checked for completeness. The visualisation helps to get an overview. If fullness is given, the causes with the highest probability can be determined to prioritise them.

The Ishikawa diagram is a beneficial tool to examine influencing factors and their sources of error. Therefore, the Ishikawa diagram is one of the seven quality tools and represents as an A3 problem-solving sheet (see Section 2.5.5) one part of the 8D[3] report (Bertagnolli 2018).

The creation of such a diagram is instrumental in group work. Many different perspectives can be integrated. However, when interpreting the chart, it must be taken into account that no interactions and time dependencies are recorded.

2.5.4 A3 Method and Report

An A3 report is a one-page report prepared on a single DIN-A3 sheet of paper (297×420 mm) that presents a problem and decision-making process in a transparent, clear, and easily understandable way. The A3 report adheres to the discipline of PDCA thinking so that the problem-solving process is conducted in a structured and collaborative way. This enables the A3 report user to illustrate complex contents simply and find practical solutions. The decision-making process is iterative to identify leading root causes (Chakravorty 2009).

It is important to note that the A3 methodology is a problem-solving strategy and not just a report. The A3 process is particularly noteworthy in that the approach

---

3 The 8D method is a problem-solving approach that originated in the automotive industry. In order to solve a problem with an unknown cause, the team works through a total of eight steps in a defined sequence and documents the result in a comprehensible way. The creation of 8D reports for the customer is a widespread procedure in quality management with the aim of sustainably eliminating errors that have occurred once.

to problem-solving is presented transparently through successive steps of analysis and action. This leads to a better understanding of the underlying problem, whereby better and more sustainable solutions can be designed and implemented (REFA AG 2021).

The following approach shows an example of how to conduct an A3 problem-solving process (see Figure 2.25).

Step 1: Background and problem statement
In the beginning, the problem is described under consideration of the boundary conditions. Numbers and other information are used to understand the context of the problem.

Step 2: Actual situation
Using diagrams of the current situation (e.g. the current process) and other graphic highlights, the problem is described in more detail regarding the status quo.

Step 3: Goal statement
In this step, a clearly defined goal is described. For instance, the plan can be described in a well-structured way using the SMART (specific, measurable, achievable, relevant, time-bound) method. All in all, it is essential to define a goal that is relevant to the problem.

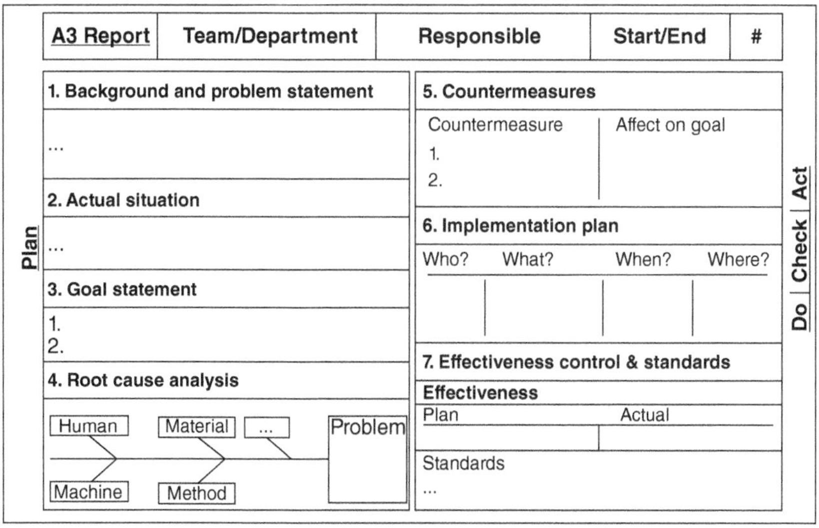

Figure 2.25 A3 report.

Step 4: Root cause analysis

With the help of tools like the Pareto diagram[4] or the Ishikawa diagram, the leading causes of the problem shall be identified. This is followed by the application 5W to determine the actual cause.

Step 5: Countermeasures

Based on the root cause analysis, measures are derived from eliminating the problem/cause. This includes checking how the new process flow is to be designed.

Step 6: Implementation plan

In the next step, the measures will be specified. Among other things, the following questions must be answered: 'Who should implement the action by when? What is to be achieved with the measures in concrete terms? In which sequence should the standards be implemented?'

Step 7: Effectiveness control and standardisation

A3 is an iterative process. The control of results is crucial for the long-term introduction of new standards. The measures introduced are compared with the planned goal to check whether the action has achieved the desired effect. If there is a deviation between target and actual effect, further countermeasures must be added. If a measure meets the requirements, the new process is established as a standard in the project.

According to Christoph Marquardt (2018), if you have carried out a large number of these A3 reports, you should also notice on the construction site that the structured problem-solving process saves time overall because you do not have to keep chasing after problems.

According to the Lean Construction Institute (2015), the benefits of using A3 are:

- Diversity and flexibility in application.
- Improved collaboration, outcomes, and the problem-solving of specific issues.
- Identifying root causes.
- Sharing information and knowledge among team members.
- Reaching consensus among team members.
- Promoting deliberative, thoughtful decision making.

2.5.5 Visual Management

Appropriate visualisations (Japanese: *mieruka*) and visual management, in general, are among the core requirements of lean management. From an evolutionary

---

4 Based on the Italian economist Vilfredo Federico Damaso Pareto (1848–1923). The Pareto diagram is based on the principle of the same name (also called the 80/20 rule), whereby it is assumed that the majority of effects (80%) are due to a smaller proportion of cases (20%). The representation takes place using a column diagram, which arranges the contained values in descending order of magnitude. The Pareto chart finds application in statistics and is used as a quality tool for error analysis.

biological point of view, the visual reception and processing of information are the fastest and most influential in humans. Owing to technical progress, however, the flood of information in the most diverse areas of daily life is increasing more and more. The addressees of this information can no longer grasp it without problems. Gorecki and Pautsch cite road traffic and the workplace as examples of this. Visual management offers standards and specifications for implementing visual representations to prevent such issues. Above all, the aim here is to convey information quickly and in the most ergonomic way possible (Gorecki and Pautsch 2013).

To be able to use visual management in such a targeted way, it is oriented to the three realities, as presented by Kostka and Kostka:

- The actual place, i.e. the confrontation with the problem at the place of occurrence.
- The real thing and thus all possible influencing factors which have an effect at the actual place.
- The fundamental knowledge is revealed by intensively questioning the employees on-site and using their knowledge (Kostka and Kostka 2013).

The various visual management tools include:

- Display boards: signboards with a combination of standard signboards along with specific colour.
- Colour coding: highlighting items with different colours (e.g. signaling status quo with a traffic light and direct escalation to project manager).
- Visual display of data in the form of graphs.
- Augmented construction field visualisation with augmented reality technology.

Besides making problems visible, visual management generally includes various other objectives:

- More transparent presentation of processes, process goals, and the current degree of fulfilment or implementation status.
- Enabling self-control and improvement of processes by employees (Kostka and Kostka 2013).
- Generating proactive interaction between employees, supervisors, and experts.
- Communicating corporate goals and values across all hierarchical levels (Gorecki and Pautsch 2013).
- Making problems and waste visible and eliminating them.
- Increase employee motivation and identification with the company and its goals.
- Documentation and support of continuous improvements (REFA AG 2022).

However, how visualisation is designed is strongly dependent on the place of presentation and the type of presentation chosen. To realise the desired success of visual management, the updating cycles and the responsibility for updating must be defined from the beginning.

Figure 2.26 Central information board in production department.

According to Dlouhy et al. (2018), visual management is essential for implementing lean thinking. It means using visual aids to control processes and transparently present facts and activities. With the help of visual management, all project participants can quickly gain an overview. The main task is to provide vital information for all project participants to discuss the same topic.

One possible application area for visual management is the design of information boards on the shopfloor. In this case, particular attention must be paid to the clear and appropriate presentation of the information.

Figure 2.26 shows an example of such a central information board in the production department of a medium-sized manufacturing company.

2.5.6 5S/5A

5S, sometimes also referred to as 5A in German, is an approach from the TPS to ensure order and cleanliness. It is one of the essential elements of lean management, in which the elimination of waste and the standardisation of processes at individual workplaces are to be realised.

In detail, these are (Gorecki and Pautsch 2013):

2.5.6.1 Seiri – Sort
Unnecessary materials and tools are banned from the direct work environment. This reduces the amount of material stored there. The subjective perception of the employees should not be disturbed by irrelevant utensils.

2.5.6.2 Seiton – Set in Order
Tools and materials are assigned a fixed position. Markers indicate the position. Ergonomic factors play a role in the choice of positioning. Tools are aligned so

that they can be taken directly without performing many manipulations. The goal is to create a kind of automatism for the user to have direct access to the tool or material.

2.5.6.3   Seiso – Shine
The workplace is cleaned from the ground up so that dirt and waste are no longer present. In the course of thorough cleaning, any existing damage to machinery and equipment can also be detected and subsequently repaired.

2.5.6.4   Seiketsu – Standardise
The new state of cleanliness must be defined as the new standard. This standard must be adhered to by all parties involved. The quality of the individual process step then benefits from this, as does the process itself.

2.5.6.5   Shitsuke – Sustain
The last step is a kind of a reminder. 5S is a cycle that always starts over again. Discipline for order and cleanliness is the essential prerequisite for optimising processes.

The goal is to organise the workplace so that efficient and effective standards, stability, and organisation are created at the place of value creation. To help employees internalise these goals, visualisations of the topic, such as the one shown in Figure 2.27, are often used to remind them of these goals, both in analogue form (e.g. by posting notices) and digitally (e.g. on the intranet).

With 5S, every employee is encouraged to improve their work processes continuously.

The method's focus is on safety, order, and cleanliness, whereby the process-oriented arrangement of objects, visualisation through markings, quality improvement through regular cleaning, and strengthening of the team spirit through the involvement of employees are used to achieve the same. In some cases, the term '6S' can be found in literature, where the sixth 'S' stands for 'shukan', which describes the intended habituation effect (Bertagnolli 2018).

2.5.7   Plus/Delta Review

A popular and universally applicable tool in lean construction projects is 'plus/delta', also known as 'pro and contra', 'do again/do better', and 'plus/change'.

The technique originated at Boeing, where it was developed in the 1980s. From there, it made its way to Greg Howell and Glenn Ballard, which led to the method quickly finding its way into the LPS and the meeting processes of the LCI and the International Group for Lean Construction (IGLC) (Mossman 2019).

The tool is particularly useful for continuous improvement. Frequent-use cases are at the end of a meeting, workshop, or activity to retrospectively record

Figure 2.27 5S method.

positive aspects and those in need of improvement. With long-term and sustainable application, existing processes can be improved step by step. The emphasis here is on step by step, as the goal is to improve the process continuously with small and continuous efforts.

Ten minutes should be planned for the plus/delta. In the beginning, it is essential to explain to the participants what plus/delta means.

According to Mossman 2019, plus/delta includes the following questions to be asked:

'Plus' means: What brought value? What do we want to keep?
'Delta' means: What can the team change or add to bring more value? How can the team do better? How specifically might we improve this process in the future?

When formulating plus/delta, it is essential that the participants refer to the process of the meeting, workshop, or activity and not to the person in charge. It is also important to encourage participants to formulate their plus and delta precisely and not too generically. If there is enough time, it is recommended that participants develop their own on Post-its and then presents them in the plenary session. During the presentation, the other participants listen. Afterwards, questions or possible solutions can be formulated.

The moderator or person responsible for the event has the task of creating an environment for the participants to feel confident to communicate their concerns. It is also important that the moderator symbolises that they will process the feedback further. An example could be to include the above points in a kaizen list to consider the feedback for the following events.

At first glance, plus/delta appears to be an easy tool. But this is precisely its strength. Project participants quickly understand how to use the tool. Furthermore, it requires no preparation and can be rapidly implemented. The lean culture helps to highlight strengths and identify areas where the action is needed.

2.5.8 Big Room

Looking at the origins of the big room, direct references can be made to Beer's 'Management Cybernetics' (1963). Stafford Beer's 'operations room' was a cybernetic interpretation of the war rooms that originated in World War II.[5] The task of these rooms was to make the same relevant information available to all decision-makers present, develop options, discuss them, and simulate them accordingly directly on the spot on the available maps (Pfiffner 2020).

Stafford Beer developed his operations room for the Cybersyn project in Chile (1971–1973), which can be seen as a blueprint for the big room. Beer's operations room was the centrepiece of the project and very futuristic by the standards of its time.

The objective of the room's design was, on the one hand, to consolidate and integrate all strategic activities and, on the other hand, to incorporate strategic knowledge and strategically relevant information into the decisions of top management (Hetzler 2008). Beer states in *The Heart of Enterprise* that the advantage of such a space is in the aspects already mentioned in this chapter and in creating an environment that best supports human decision-making, which is subject

---

5 Based on the principles of cybernetics, Stafford Beer developed management cybernetics in the late 1950s. This describes complex social systems as viable entities and offers approaches for planning and controlling them. Beer developed the viable system model, which is intended to ensure the controllability of a system despite high complexity, dynamics, and limited predictability.

to neurophysiological limitations (Beer 1979). Further detailed considerations on this topic can be found in Hetzler's 2008 dissertation 'Brain supporting environments for decisions in complex systems'.

A big room brings together employees of various planning parts and client representatives, project managers, and employees of general contractors, subcontractors, and leading suppliers for the project duration. The main goal is to increase collaboration significantly and thus communication within the project. The big room helps to highlight the interface problems by placing all participants in an open-plan office. In addition, the big room aims to reduce silo thinking among the project participants and develop a common project thinking. The big room is a helpful tool for integrated project delivery (Verein Deutscher Ingenieure 2019).

A big room is designed so that collaboration between project participants is made possible. Therefore, the systems used must harmonise with each other and meet the users' requirements. For example, information exchange formats must be defined in advance. In addition, the room is equipped with tables, writable surfaces, screens, and other utensils that enable joint work. Besides analogue media, digital infrastructure is required in the big room. Internet connection (cable and wi-fi) aside, the integration and visualisation of digital building information models can increase collaboration in the big room.

However, the big room design is not finished with the equipment. It is essential to direct the project participants' behaviour to guarantee a partnership-based working method. Therefore, it is helpful to define rules of play for meetings and the overall interaction in the big room. The interaction between those involved in construction in the big room should be analysed, and, if necessary, measures to promote cooperation should be initiated. A helpful approach to ensure this is the introduction of pull planning (LPS), in which all processes of a construction project are described to provide a global view of all subprocesses. Milestones are set, and backward-looking planning is used to develop a schedule.

According to Ron Cruikshank (2019), an author on http://Leanconstructionblog .com, the key elements to creating a practical big room are:

- Behaviours and rules are clearly defined.
- Together, the project team defines the conditions of satisfaction.
- An environment with 'respect for people'.
- Open and transparent discussions are supported.
- Commitments are made and kept.
- Problems are addressed proactively, in a collaborative manner.
- Breakdowns are declared, with no blame or finger-pointing.

2.6   Practice Insights from Martin Jäntschke

2.6.1   Infrastructure Railway – Introduction of Lean Construction in Large Projects

The following project example for the introduction of lean construction in large projects comes from the railway infrastructure sector. Here, large-scale projects are understood to be projects with a total value of more than €50 million. This budget limit is intended to serve as a boundary to a certain complexity. Railway infrastructure projects are generally characterised by a high number of different disciplines. They are usually the subject of exposed public discussion, and subject to a high degree of dynamism (such as political will, public perception, climate change, taxpayers' money, etc.) from the external boundary conditions, perhaps even somewhat more than other major projects.

The chosen example is a line project with about 85 line kilometres. The line length is divided into several sections (so-called planning approval sections) – in this example, there are seven of them.

In addition to the 85 km of superstructure and alignment work, the project includes 37 bridge structures, various culverts, and four stops or stations to be upgraded to the 'PRM' (people with reduced mobility) standard. In addition, two new electronic signal boxes (ESTWs) for train control and communication are to be built here. Of the trades involved, structural engineering, superstructure, technical equipment, and electrical energy systems are primarily mentioned here. Each specialist contract is assigned to different planning offices with separate tenders and contracts for each planning approval section (PFA). In addition, external experts have been commissioned for subsoil, noise, and vibration protection; environment; and electrical emissions.

Further to increasing the existing line speed from 160 to 200 km/h (travel time gain and utilisation of the line), the project objective also includes electrification (construction of a new overhead line system), renewal of the control and safety technology (LST) to the latest standard and upgrading the stopping points to the 'barrier-free' standard.

Therefore, the complexity and dynamics described here are definitely present in this project example.

The measures and tools of lean construction were only included in the course of the tender for the construction work – the entire planning process was carried out in a 'classic' (sequential) manner. Text modules of the ZIB initiative (Future Initiative for Railway Construction from Deutsche Bahn), which were developed by the railway infrastructure companies and the construction industry, were included in the tender.

The instruments described in these text modules are as follows:

- Establishment of a 'lean room' (big room).
- Use of the LPS.
- Use of the A3 report for decision-making needs.
- Notes on the meeting system.

The introduction of these tendered and contracted elements was implemented in three project phases:

1) **Creation/plausibility check of a cross-trade construction schedule**
   The first phase was characterised by creating a cross-trade construction schedule and the plausibility and combination of the existing construction schedules of the individual contractors. In the process, cross-trade process overviews were produced in which up to 250 unique processes were included. Twenty-five project participants achieved a common understanding instead of the individual contract objectives. The existing construction time planning was checked for plausibility and expanded as needed. A comparison was also made with the already registered closure times, and, if necessary, these were adjusted.

2) **Implementation of a joint task and risk management**
   The planning and execution of the individual trades were secured through joint risk management and the client's participation. The contractors' personal risks were summarised, identified, and/or defined. A standard task management system was also introduced to create transparency across all contracts, to discuss the identified risks, and to determine joint countermeasures. This form of structured, cross-trade processing and the short-cycle tracking of risks ensure high project stability during the realisation phase.

3) **Training of project participants and knowledge transfer**
   The early involvement of all those responsible for implementation ensured the sustainable anchoring of the approach. The transfer of knowledge and creating a standard 'lean ABC' was provided through several workshops with different participants – from the senior site manager to the foreman. Here, the focus is on continuing the methods and their sustainable anchoring in the minds of all those involved in the project.

These three project phases are described in more detail and concretised in the following with the individual project results.

1) **Preparation of process overview – trade sequence**
   Each trade presents its schedule with the contractually agreed framework dates, including the rough sequences and associated dependencies. Subsequently, these individual sequences of all trades are superimposed, and thus

an overall independent overview in chronological order is created. As a result, a cross-trade perspective is created – the goal of an individual trade is no longer in focus – the common project goal is what counts. This can result in individual trades not working in the optimal sequence for them or resources (personnel and machine deployment) shifting within the solitary schedule. However, it can also happen that the contract deadlines that were initially set turn out not to be target-oriented and thus have to be adjusted.

2) **Comparison and supplementation based on the consolidated schedule**
This reconciliation creates transparency across all trades and represents detailed coordinated planning. The existing general schedule is supplemented and, if necessary, adjusted by the sequence of trades and updated over the entire construction period. At the same time, this detailed schedule provides a reliable basis for site control. Since all trades are represented here with their sequences, it is also essential on the client's side to make any contractual adjustments that may be necessary with the respective contractors.

3) **Task and risk identification based on existing trade sequence**
Since all intermediate deadlines are now coordinated and synchronised based on the detailed schedule and the required fulfilment criteria and using sequences and dependencies, the risks that arise can be recorded in a comprehensive and structured manner. Based on this overview, the assignment and corresponding linking of the possible dangers to individual process steps can be easily carried out. Through the open and transparent presentation of risks, a problem-solving culture is initialised, risks are jointly anticipated, and possible countermeasures are developed together.

4) **Task and risk identification**
Clarity must be achieved on the steps required to minimise risks. The listed risks are entered in the process overview and thus visualised for all involved. This promotes an overview of the remaining steps to safeguard the planned measure. Tasks resulting from this are transferred to the summary and later broken down into correspondingly required subtasks. In this way, the target dates with all their necessary fulfilment criteria are jointly agreed among all project participants.

5) **Task management**
As the last and hierarchically lowest step, all tasks from all previous actions are now transferred to a task management system. It is advisable to use appropriate software for all projects. This software should virtually support the transparency of the pending tasks. The task managers' commitment is necessary since everyone does not always desire the given transparency – here, the client can set a good example. Things that still need to be clarified or escalations are definitely in the client's sphere. The project's overall progress is jointly evaluated through regular follow-up tasks, including (re)prioritisation and synchronisation.

Owing to the increasing limitations of visualisation because of many individual tasks and risks, it was decided in the project to reduce to so-called focus areas for the time being. These were selected based on the number of assigned processes and tasks examined in greater depth. These focus areas could be easily 'read' simply through the transparent and cross-trade visualisation of the overall project.

The existing risks and tasks were then transferred into subtasks and corresponding categories based on their responsibilities in the focus areas. Subsequently, a qualitative assessment is carried out here. To counteract a possible selective task load, these subtasks were included in the regular deadline of the task management. Here, the subtasks were evaluated about the target date to be met and, if necessary, reprioritised. The task load of individual subarea levels (e.g. trade or project partner) could thus also be easily read and counteracted with foresight. For the subtasks of the focus buildings, a degree of completion was defined for each calendar week as a unit of measurement, which was set about the actual forecast degree of completion. In this way, it was possible to determine the progress of the tasks and possible over- or under-fulfilment levels.

The elements of advanced planning and target/actual comparison were introduced based on the jointly prepared construction schedule to control the construction site areas. The detailed scheduling across all trades forms the basis for the target/real comparison of the construction site control.

The preliminary planning is regarded as a binding fine-tuning of all participants and is stable with a detailing horizon of 8–10 weeks. All project participants chose the 8–10 weeks specified here – these take into account materialisation and work in preparation for construction by the contractors and approval/legal and internal company processes at the client.

The target/actual comparison is always carried out on a short-cycle basis and serves to control and harmonise all capacities available on the construction site. In addition, it is the instrument with which target deviations are to be recognised at an early stage and with the help of which any countermeasures can be derived. The comparison offers the first step of the ZIB – the collected experience values are conserved here and transferred to the scheduling or advance planning.

2.6.2 Implementing Change in an Infrastructure Organisation

Implementing change in a company structure is classically carried out through various measures (waterfall). Culture development as a programme, operational excellence (OPEX), the introduction of agile working methods, or the introduction of 'lean management' methods are the usual examples here.

But isn't it the case that, despite the many programmes, the sustainable effectiveness of these measures is very often missed?

Typical questions and issues arising from this 'programme culture' are often 'old wine in new bottles', 'another pig in the poke', 'when will this (finally) end', 'what does this do for my daily work'… and many more. This can conclude that these widespread approaches, which are supposed to bring improvement, do not work sustainably, at least not in this way. But what could be the reasons for this?

Using a practical example that all of us know, I would like to show a possibly more sustainable and goal-oriented alternative to the previous models. The example is about pushing an already swinging swing while blindfolded.

The swinging swing is supposed to represent the already existing organisation. The swing swings – the organisation moves – in this case, even rhythmically. The organisation is moving, that is it is alive, and it seems to be functioning – because it is swinging (moving). It is, therefore, 'successful'. But over time, the momentum is lost if the swing is not pushed, whether from the outside or when one sits on the swing oneself (i.e. anchored in the organisation). If the swing stops, the organisation is also motionless, so to speak: it is no longer successful. The change process I would like to show with this example would be the case that someone from outside wants to push the swing blindfolded. This could be compared with (mostly) external consultants who do not know the organisation exactly (not sitting on the swing) and are supposed to make a change here (blindfolded). So what is the best way to do this in this example?

First, you have to find your way around blindfolded. You have to move towards the swing without coming into your path (otherwise, you will be hit unintentionally). You have to move to the right place to bump (often with small steps). Then, when you feel you are in the right place to bump, you should feel the swing swinging. Of course, you can take off your blindfold in between to make sure you are on the right, purposeful path. Then you should push the swing at the right moment. If you do it too early or too late, the swing will lose its swing. The impulse of the push should also be well-considered: if the push is too light, nothing will happen in the swing; if it is too violent, the swing will swing too much and hit the person making the push at the next amplitude.

I think you can already see what this example is alluding to. The example shows that many organisations are successful at their core and that the 'change' push does not have to be so great (too strong a swing). Or that the 'change' push does not come at the right time (amplitude of the swing) and thus does not move the organisation forward but instead slows it down. It is also possible to deliberately slow down and interrupt the swing – but this should be urgently communicated beforehand. Otherwise, the person sitting on the swing will fall off. This means that the 'change' should work with the organisation, have specific connectivity (adapted to the organisation), the cultural patterns should be grasped and felt (swinging of the swing), and the goal of the 'change' (the impulse of the push) should be well-dosed so as not to irritate the swinging of the swing too much. In the best case, there is always a 'little change' – in our example, the swinging person always swings a little so that the amplitude always remains the same – against gravity. Then it would only be noticed that no change occurs when it is no longer made – in our case, the swing slowly swings out.

From my point of view, this example shows very clearly how sustainable change can look effective and positive for the company:

- Change should be connectable (frequency of the swing).
- Deliberate irritation of the organisational system (the impetus is deliberately provided from the outside).
- Working *with* the organisation (working with the swing of the swing).
- Change is constantly occurring within the organisation (the person swinging is always swinging a little themselves).

The following real-life example is intended to illustrate this thesis (swing) once again.

In 2018, several 'strategic initiatives' were launched in the railway infrastructure organisation concerned, which were waterfalled down and rolled out into various detailed programmes, some of which are described elsewhere in this chapter. Part of these initiatives was also the approach to massively accelerate the organisation's business processes 'Project Management in Major Projects' and reduce the overall lead time of infrastructure projects by up to 25%. With an average project duration of 20 years, we are talking about five years as a pure acceleration effect here. In addition, an increase in the quality of execution and more efficient use of closure periods for the execution of construction work were given as further goals of this initiative.

In the course of the programme 'Business Process Excellence (GPEX)', the projects were asked about the 'time killers' in their opinion, and feedback and examples were obtained at all levels.

The feedback ranged from very general statements such as 'too few resources', 'political environment', 'duration of additional work', and 'poor performance by third parties' to concrete examples in project implementation, e.g. 'the draft resolution XY of XX.XX.201X was rejected four times by the board's anteroom due to "deficiencies" and the originally submitted version was finally submitted for signature after eight months'.

All these 'time wasters' reported back were sorted according to content and packed into topic clusters. The clusters were then entered into a 'cause/effect' overview matrix. The 'quick wins' and the topics with a high impact were selected and processed further in the following.

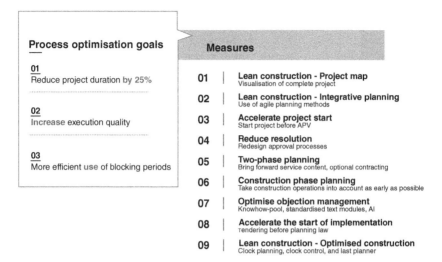

Process optimisation goals

01
Reduce project duration by 25%

02
Increase execution quality

03
More efficient use of blocking periods

Measures

01 | **Lean construction - Project map**
Visualisation of complete project

02 | **Lean construction - Integrative planning**
Use of agile planning methods

03 | **Accelerate project start**
Start project before APV

04 | **Reduce resolution**
Redesign approval processes

05 | **Two-phase planning**
Bring forward service content, optional contracting

06 | **Construction phase planning**
Take construction operations into account as early as possible

07 | **Optimise objection management**
Knowhow-pool, standardised text modules, AI

08 | **Accelerate the start of implementation**
Tendering before planning law

09 | **Lean construction - Optimised construction**
Clock planning, clock control, and last planner

Figure 2.28   Project optimisation by measures and methodological approaches.

As a result of this pre-sorting, six process measures ('quick wins') and three methodological approaches were pursued as respective goals (see Figure 2.28).

The three methodological approaches are listed here as 'Lean construction' measures (changing the way of working).

The process measures can be introduced into schedules or the flow of existing processes as 'mechanics' according to a relatively simple principle and have an immediate effect, so to speak 'quick wins'. The improvement for the employees (acceleration effect) is immediate: in the best case, the employees ask themselves, 'Why haven't we always done it this way?' This is a clear indication of the hygiene of value creation and that the measures taken were chosen correctly in terms of their effect. It is more time-consuming and ultimately not as measurable with the changed methodical working methods. Here there are no mechanics, the effect does not unfold immediately, and, finally, the structures around the employees have to be changed. Important: it is not the employees who are changed but the structure around them! Last but not least, the measures cannot be directly converted into an accelerating effect, so the management must remain patient in order not to jeopardise the impact of the steps.

To ensure that the methodical working methods can be introduced and anchored in the organisation sustainably and effectively, a team of so-called lean coaches was established for this purpose. They introduce and accompany these measures and working methods together with the project teams.

This introduction proceeded as follows:

1) **Initial meeting with the project manager**

   Here, the project was presented, the expectations of the measures were collected, and the project-related specifics were discussed. In addition, the relevant measures and dates were agreed upon.

2) **Preparation of the workshop series**

   The organisation of the dates, the premises, and a 'readiness' check before each workshop were clarified here and prepared accordingly.

3) **Migration workshops**

   The relevant measures for the project teams that had been classified as goal-oriented in the initial discussion were presented here in project-specific so-called modules in workshops and practised in a practice-oriented manner (on the respective project).

4) **Accompaniment**

   Untypical of process measures or centrally distributed measures in general, the project teams were accompanied by 'lean coaches', meaning the project teams always had the same contact person for questions or improvements. Depending on the project team, the measures were followed up individually and, if necessary, supported with a further series of workshops (if deemed necessary).

5) **Assessment**

   After the measures have been successfully taken over, a so-called assessment, ordered by the project team, takes place. In this assessment, feedback is obtained from the team, the success of the measures is evaluated, and improvement measures are introduced if necessary. The assessment is the first step toward a culture of improvement and thus strengthens acceptance.

All in all, we can proudly state that the organisation of lean coaches with a strength of eight regional and three central lean coaches has introduced these measures and working methods in over 110 major projects all over Germany within just under two years. The feedback from the project teams is consistently positive – so positive that the 'smaller' infrastructure projects are also claiming these measures for themselves and want to introduce these workshop series as well. As a countable success, it can be stated that the 11 lean coaches have achieved an acceleration effect of over 850 person-years in the projects in two years (i.e. 22 person-years) – this already excludes any countereffects that may have taken place anyway, so that one can speak of real acceleration here.

2.6.3 Conclusion

2.6.3.1 To Section 2.6.1

Through this collaborative approach, the cooperation in the project could be significantly improved across all contractors and lot boundaries. The project goal is in focus and not the contractually agreed goal of each contractor. However, this cannot be presented in the upstream planning or described clearly and unambiguously in any tender texts. Here, the unique know-how of each contractor is decisive. Thanks to the transparency and regular, systematic follow-up in task management, potential project risks were identified early, followed up in a structured manner, and often an overarching technical solution was found. Individual emerging risks were not even escalated to the client but solved beforehand through communication between the contractors involved. For all this, a joint commitment among all project participants is essential. There were and still are supplements and/or additional services – but always focusing on the project goal. Only one software should be used for task management. It facilitates regular synchronisation and avoids unnecessary double or multiple maintenances by all participants. Through the early involvement of those responsible for implementation (management level) and the closely managed exchange, these methods could be sustainably anchored in project organisation. And last but not least, the visualisation helps focus on the relevant and high-priority issues to come a significant step closer to the common goal: a trouble-free construction process.

2.6.3.2   To Section 2.6.2

Changes take place constantly in a company. The effectiveness of these changes cannot always be measured immediately and require a specific 'protective space'. 'Change' should always refer to concrete processes or examples – only in this way can they be solved concretely. In my opinion, the example with the swing illustrates very well the method in which change (impetus) has to take place. Sometimes you need the impulse from the outside: in the best-case scenario, the person swinging always pushes a little themselves, in other words away from programme culture and towards improvement culture!

2.7   Findings, Criticism, and Reflective Questions

2.7.1   Findings

Some of the approaches to lean that are considered fundamental today actually date back several centuries. This could lead us to conclude that the optimisation and simplification of work and production processes have always been the focus of respective users. Lean management, with its roots in the TPS, results from a shortage economy. For this very reason, lean has lost none of its topicality about avoiding waste and the sustainable management of scarce resources.

In addition to the tools of the TPS, the five central pillars of kaizen: process, customer, quality, criticality orientation, and standardisation are essential guidelines that are equally important for all industries.

At the heart of lean management, which emerged from the TPS, are countless general applications. Still, methods and tools, particularly the four principles, can be adapted for almost all other industries: the demand for flow, takt, pull, and zero defects.

Lean construction is the adaptation of lean management to the construction industry. Through a process-oriented and cooperative way of thinking, unnecessary waste can be avoided or transformed into added value. This shows that, based on the general requirements of lean management, particular methods and tools have also developed over time for the construction industry, as various methods to analyse processes, establish cleanliness and organisation, and create continuous improvement of processes. Lean construction has established itself as a new way of thinking supported by systems, methods, and tools.

Nevertheless, it should be noted that many of the lean construction methods and tools mentioned in this book are used in the same or similar way in other industries.

An example of this is the practice insight from Martin Jäntschke. In his description of an implementation project of lean construction in the railway infrastructure sector, the author comes to the following central findings regarding the design of intelligent construction projects:

- The project goal, and not the individual goals, is in focus. All project partners must be interested in the project, take responsibility accordingly, and not think in terms of economic units.
- Change is about value hygiene – not social hygiene – which can also lead to 'defensive reactions' in the company. Sustainable improvement has only taken place when it is noticed and no longer done.
- Examples:
  - Don't want to change everything at once; always observe the reaction of the organisation.
  - Don't want to change the employees but the structure around them.
  - Find a good mix between quick successes and longer-term measures.

2.7.2 Criticism

In addition to numerous publications that highlight lean management, its industry-specific variations, and their positive effects, as well as countless companies that emulate the example of Toyota Motor Corporation in their search for greater efficiency, there is equally extensive, and in some cases justified, criticism of the topic of lean.

'Lean Management is a method that the Europeans use because the Americans think that the Japanese use it successfully' is the somewhat deliberately ironic statement that H.J. Klepzig (2018) prefaces in *Lean Management in Practice*.

Klepzig cites the following specific criticisms of lean, which can also be found in other critical sources on the subject:

- Lean is not a self-contained strategy but rather a "toolbox".
- Various requirements of lean would be in contradiction to traditional, Western corporate management and would therefore tend to be unsuitable, or at least difficult to implement, e.g.:
  - The call for continuous improvement would not correspond to the Western way of thinking we are familiar with.
  - Avoidance of waste is not automatically equivalent to (more) value creation.
  - Numerous lean tools promote inflexibility through rigid requirements, which limit the company's ability to respond to the market and are therefore counterproductive.
  - The Western production philosophy often contradicts the requirements of lean in various areas (e.g. layout) and can therefore only be implemented with difficulty or inadequately.
- In general, there are often deficiencies in the introduction and the subsequent anchoring in the company.
- Statements about possible successes of lean can often only be made to a minimal extent, as these are usually not quantifiable at all or only in fragments, owing to a lack of measurements (Klepzig 2018).

Furthermore, among the frequently mentioned criticisms regarding lean management are the following:

- General problems with the adaptation of lean due to cultural differences to the country of origin, Japan.
- Overemphasis on the efficiency approach of lean in Western culture (up to the rationalisation of employees).
- Failure of lean due to lack of 'fast' successes.
- Minimisation of stock and personnel can lead to bottlenecks in case of disruptions.
- Focus is only on the elimination of wastage.
- Burden on employees due to constant monitoring.
- Increased pressure to perform due to zero-defect requirement.
- Lack of clarity due to a large number of tools.
- Success is strongly dependent on the acceptance of the employees.

Just as lean construction has particular methods and tools tailored to the construction industry, the literature on this subject contains both critics and defenders. As only two prominent examples, the second edition of Graham M. Winch's *Managing Construction Projects* (2009) should be mentioned on the side of the critics. And as a direct reaction to this, 'A Response to Critics of Lean Construction' by Ballard and Koskela (2011).

Even though lean construction brings countless added values, the introduction of lean is often a big challenge, which often fails. Positive effects only emerge when project participants get involved in lean construction and are open to change. Change management is a decisive factor, which engineers often ignore. This is fatal for the introduction of new methods. In addition, the introduction of lean needs to receive support and assistance from production (bottom-up) and general management (top-down). In summary, the application of lean construction is only successful if it is done together.

A further point of criticism of lean construction is that many applications and levers work as a matter of course for some people. In other words, lean is logical common sense. Especially for companies that are collaborative and process-oriented by nature, they will experience less added value than an unorganised company. Instead, the already efficient company will have to subordinate itself to the structures of lean construction because the other project participants require it to do so.

Another point of criticism is that certain parties use the collaborative approach to exploit the other project participants.

Martin Jäntschke was also able to identify some points of criticism in the context of implementation in the practical example he cited:

- The existing tender texts do not help in real work, as the methods and commitment cannot be incorporated into the contract for all.
- Sometimes an impulse from outside is needed; protective spaces are sometimes difficult to create for leaders, as quick successes are desired.
- Copying or adapting positive examples is not possible, as the organisation's respective context (up to the employees) must always be considered.

2.7.3 Reflective Questions

- What approaches could be used to perfect the pursuit of efficiency without dragging employees down in the process?
- What does the commitment to lean thinking mean in dealing with employees?
- How does the corporate culture change in the context of lean thinking?
- How should you proceed to define the optimal lean method mix for a particular company or project?
- In which projects would you prefer, LPS or takt planning and control?
- Do you think there are further types of waste in construction projects?
- Would you prefer collaboration or control? Please explain why?
- Lean construction originated in the 1950s. Do you think it is obsolete in the current time?
- Do you think lean thinking generates a more significant value in the design or implementation phase?

3

Cybernetics and Lean

When something is important enough, you do it even if the odds are not in your favour.

Elon Musk

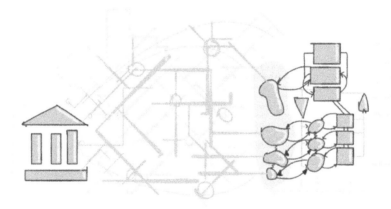

Before various approaches to linking cybernetics and lean are examined in more detail in this chapter, first comparable research approaches are briefly discussed, to show that a general connection between cybernetics and lean is reasonable, purposeful, and fully justified.

Initially, we should refer to Jaqueline Sharma's synthesis of scrum[1] and Malik's management models.[2] Sharma attempts to link the two approaches based on the identified synergistic effects to manage complexity and create positive differentiation factors for companies (Sharma 2012). Although scrum can be seen as an implementation of lean development, it is not in Sharma's primary interest to use the total efficiency of lean. Also, although Malik's model is based on Stafford Beer's viable system model (VSM), it is a variation.

Brecher et al. (2015), in their considerations of "cybernetic approaches in production engineering", deal primarily with the technical aspects of cybernetics; however, they also state in their conclusion that in addition to focusing on a production process, a higher-level synchronisation of the processes must be aimed for, which can only be accomplished by self-organisation. This is the fundamental basis for an adequate response to changes in boundary conditions.

In 2011, Herrera et al. demonstrated at the International Conference on Industrial Engineering and System Management that the VSM was suitable for designing product-oriented systems, especially when a trade-off between control and task anatomy needs to be found. Herrera et al. use intelligent production systems as a connecting element.

While the topic of lean has only been touched on in passing in the publications presented so far, Herrmann et al. (2008) explicitly deal with both elements – lean and the VSM – in their conference report 'Lean Production System Design from the Perspective of the Viable System Model'. Herrmann et al. also use the VSM as a structuring element. Still, in their research, in addition to VSM-based diagnosis, they focus on considering the extent to which lean methods can be used concerning the attenuation or amplification of variety. Here, the worker becomes an extension of management through their variety (Herrmann et al. 2008).

---

1 Scrum is an agile process model for software development. It was initially assumed that the process of software projects could not be planned in detail in advance due to its complexity. Therefore, the planning is refined step by step. In addition to numerous other influences, scrum was also shaped by the 3M model (muda, mura, muri) and used several specific techniques and procedures. In the meantime, the method is applied in numerous other disciplines, e.g. scrum implements lean development for project management.
2 Fredmund Malik is an Austrian economist (*1944). Since his studies at the University of St. Gallen, his research has focused on management theory. He initially played a decisive role in developing the St. Gallen management model (SGMM). Later, Malik founded a management consulting company in St. Gallen. Among other things, Malik uses systems theory and cybernetic approaches (notably the VSM) to analyse and design management systems. Malik's management system consists of three basic models: the Malik general management model, the Malik standard model of effectiveness, and the Malik integrated management system.

In their reflections on 'Management Cybernetics as a Theoretical Basis for Lean Construction Thinking', Steinhäusser et al. 2015 also use the VSM but as a theoretical basis for presenting lean construction, i.e. the adaptation of lean management in construction, in sharper terms, and anchoring it in thinking in the long term.

Dominici and Mielke (2010) take a relatively similar approach by combining the VSM with lean. Their publication 'A Viable System View of the Japanese Lean Production System' uses the VSM as a diagnostic tool. Interestingly, Dominici and Mielke also find overlaps between lean and the VSM and state that the VSM makes the architecture and dynamics of a company visible and contributes to the understanding of lean. In contrast to the authors' considerations, the focus is not on combining both elements but on the analysis of the economic environment to successfully apply lean.

For completeness, we should also refer to the research of Schwaninger and Scheef (2016). Their empirical study 'A Test of the Viable System Model: Theoretical Claim vs. Empirical Evidence' uses quantitative research methods to demonstrate that the VSM is both a reliable diagnostic and a design tool for organisations to strengthen their viability, resilience, and development potential.

Finally, as a transition to the following mappings, consider Moore et al. (2015). They combine the VSM with the Toyota Production System (TPS) with the help of knowledge management to create a sustainable, holistic management model. Moore et al. also state that combining different factors is necessary to account for complexity. Furthermore, they concede that lean only takes the design of the organisational structure into account to a limited extent. However, their approach explicitly refers to service organisations in the nonprofit sector.

3.1 VSM and Lean (Construction) Thinking

Lean management has its origins in the TPS. As the term suggests, it is not just a method but a system in which everything is connected and improvements are made continuously and in small steps day after day. Niklas Modig and Par Ahlstrom from Stockholm Business School write in their 2017 bestseller *This Is Lean* how important it is to integrate one's values, principles, methods, tools, and activities holistically into a company. It is not about dogmatic approaches or pure copying of 'methods'; moreover, it is about intelligent adaptation to one's boundary conditions.

Liker and Meier (2006) are also convinced that the success of lean depends on a few essential factors. The first point they mention is: 'A focus on understanding the concepts that support the philosophies of lean, strategies for implementation, and the effective use of lean methodologies, rather than focusing on the mindless application of lean tools.' In their work *Lean Thinking*, Womack and Jones (1996) describe an approach that further emphasises the advantages of this method for the individual and society and simultaneously creates a new way of thinking about the design of human work, making them one of the creators of this technical term.

The relationship between lean thinking, lean management, and lean enterprise can also be derived from these explanations. The first stands for the fundamental theory, while lean management represents the practical application. The concept of lean enterprise goes even further, symbolising a North Star orientation about lean.

The objective of this philosophy is that lean thinking 'shows a way to achieve more and more with less and less' (Womack and Jones 1996), whereby the employee's job satisfaction is to be increased by direct feedback.

As mentioned above, cybernetics is a systems science that is somewhat older than lean and sees 'everything connected with everything' both inside and outside the system. Thus, by applying management cybernetics, a company is connected to its environment to react adequately and proactively to changing external requirements and technologies, and volatile markets and therefore ensure long-term corporate success.

This shows that management cybernetics and lean thinking serve two sides of the same claim.

Thus, management cybernetics with its rather strategic orientation – following the quotation of Peter Drucker[3] – stands for 'doing the right things' = effectiveness. At the same time, lean thinking, with its tools and methods, represents the tactical/operational design of 'doing the things right' = efficiency.

It is, therefore, apparent to establish the connection between cybernetics or management cybernetics and lean construction as mainly applied branches of the above-mentioned systemic approaches.

3.2 Mapping the Viable System Model with Lean Construction Methods

Professor Iris D. Tommelein from the University of Berkley is researching management cybernetics and lean construction (Steinhäusser et al. 2015). She concludes that lean construction tools, such as the last planner system (LPS), can be implemented more successfully in existing organisational structures using methods of management cybernetics. With the help of the VSM as a reference and analysis tool, a transparent implementation into complex structures is possible. In her work, she argues for applying management cybernetics and the VSM in production to implement intelligent organisational design. Furthermore, she argues that

---

3 Peter Drucker (1977): 'Management is doing things right; leadership is doing the right things.'

the efficient interaction of the '1-systems' in particular must be more firmly established (see Section 1.2.1).

This means more precisely:

- Increase of flow efficiency.
- Improving communication and culture.
- Reduction of the silo mentality to a cross-linked structure.

Fatos Elezi and Tobias Steinhäuser from the Faculty of Mechanical Engineering at the Technical University of Munich have dealt with this topic and found that management cybernetics can be regarded as a theoretical foundation for lean construction (Steinhäusser et al. 2015). The cybernetics approach is classified as a first-order principle according to Elezi and Steinhäuser. First-order principles are absolute truths or laws of nature that others accept and acknowledge. They have a higher degree of abstraction and are universally valid. According to them, lean construction is the complementary good and a second-order principle. Second-order principles formulate instructions for execution and contain statements such as: 'You should … to achieve this.' In contrast to first-order principles, they have a higher practical relevance.

As proof, the two have used the VSM as proposed by Tommelein as a reference and analysis model for implementing the LPS (Steinhäusser et al. 2015). Based on this, the following mapping was carried out.

3.2.1 Mapping VSM and the Last Planner System

- **S5 – Lean principles:** System 5 embodies the fundamental principles of lean thinking. The principles serve as a guideline for dealing with other project participants and the joint development of a process plan.
- **S4 – Milestone planning and collaborative planning:** System 4 is the long-term goal of the system, here the completion of the project. Considering project environmental factors, the strategic milestones for the project are determined. Based on this, a cooperative phase schedule is created with all project participants.
- **S3 – Making ready:** System 3 has the task to enable the 1-systems to work, i.e. allocation and control of resources and troubleshooting, to ensure the prerequisites for a successful implementation. S3 concretises the schedule from S4.
- **S3* – Monitoring for trust:** System 3* creates an additional viewing angle. With go gemba (e.g. site inspection) or sporadic audits, further information and a more holistic picture of the efficiency and effectiveness of production and the system can be obtained.
- **S2 – Weekly and daily planning/last planner meetings:** System 2 is used for control and is where self-organisation takes place. This is implemented through the accurate daily determination of target and actual performance, last planner meetings with short-cycle coordination, and measurement of the PPC

(planned per cent complete) value. From this, control options can be derived by the 1-systems or by S3.

- **S1 – Measuring, learning, and continuous improvement:** In addition to the execution, system 1 has the task of passing on feedback on the existing process to S3 to optimise production, and all statements made can be complied with.

3.2.2 Mapping VSM and Takt Planning and Control

Following the example of Elezi and Steinhäuser, a mapping can be conducted with takt planning and control as well (Steinhäusser et al. 2015).

- **S5 – Lean principles:** The principles of lean thinking are assigned to system 5 as described in Section 1.3.1. These are the basis for functioning cycle planning.
- **S4 – Define milestones and work content:** In system 4, the milestones of the construction project are determined and defined. Furthermore, the building structure is divided into repeatable elements, and all tasks to be performed are determined.
- **S3 – Final takt planning:** Takt planning is completed in system 3. The duration and intensity of the various work packages and the trade sequences are specified to create a detailed process description for the 1-systems.
- **S3* – Monitoring for trust:** As previously described for mapping VSM and the LPS. System 3* can be used to create an additional viewing angle.
- **S2 – Takt control board and meetings:** The clock control panel provides the various operating systems with an overview of the overall process. Information can be transmitted via key figures or status indicators. Clock control meetings are used for the short-cycle coordination of the system 1s.
- **S1 – Measuring, learning, and continuous improvement:** As previously described for 'mapping VSM and the LPS'. System 1 produces and gives feedback to S3 on the existing boundary conditions.

3.2.3 Mapping Information Channels and Lean Construction

The following specifications are examples for mapping LPS with takt planning and control.

- **Information Channel 1 – Project manual:** A generally valid project manual is known to all project participants and contains lean principles.
- **Information Channel 2 – Phase planning and clock planning:** Process planning regulates the demand and control of resources (machines, personnel, etc.).
- **Information Channel 3 – Lean culture:** A culture that understands trust and cooperation as essential components of project management.

- **Information Channel 4 – Steering groups:** Scheduled meetings with stakeholders to inform or escalate.
- **Information Channel 5 – Last planner meetings and clock control panel:** Information Channel 5 corresponds to S2.
- **Information Channel 6 – Monitoring for trust:** Information Channel 6 corresponds to S3*.

The above mapping, both for the LPS and the takt planning and control, does not claim to be complete. In numerous discussions, the authors have found that the mapping is strongly influenced by individuals' perspective so that representational statements are 'suggestions' that are welcome to stimulate discussion. But they are certainly not the ultimate truth.

Irrespective of this, the various feedback loops and corresponding communication channels in the LPS or takt planning and control for a recursion level become clear with the mappings carried out above. This suggests that, owing to the self-similarity of the VSM, the lean approach can be uniformly mapped across all participating organisations and all recursion levels. This applies to the LPS and takt planning and control.

The consideration of process and organisation is essential for successfully implementing management methods and the strategic orientation towards an organisation that systemically integrates lean construction.

3.3 Mapping the Viable System Model with Lean Management Methods

The following excursus includes mapping the VSM with lean management methods and tools to round off the topic. The starting point for the considerations is a plant and mechanical engineering application.

This mapping aims to approximate a management system that includes the effective control of the company while maintaining a high level of complexity, principles, and tools that focus on the efficiency of the processes in the company. Therefore, the VSM is selected as the structure concretised in the company's everyday life with very operationally oriented methods of lean management (Roll 2018).

This selection is justified by the central examination of cybernetics with complexity and dynamics, which could prove to be a suitable model for the above industry- and company-specific problems.

This also benefits the method of lean management, which many companies usually understand as a loose collection of efficiency-increasing methods. By integrating it into the structure of the VSM, a sustainable management system can be generated in this way, which can also adapt the corporate strategy to the constantly changing requirements of the corporate environment.

According to Stafford Beer, the primary goal of this mapping is to use the very abstract, generally valid approaches of the VSM to anchor the operationally oriented lean methods and tools in a management system sustainably. At the same time, the application of lean tools serves to operationalise and model the VSM.

Finally, when designing a company-specific organisational model based on the VSM, special attention needs to be paid to the strategic connection of all departments or systems to ensure that all employees identify with the values and goals of the company.

Furthermore, a cultural change in the company about self-organisation and the application of lean management is aimed for, which is to be brought about by increased involvement of the employees at the operational level, and associated managers. They're being equipped with further-reaching competencies and areas of responsibility (Figures 3.1 and 3.2).

In the long term, the company expects the parallel application of VSM and lean as an organisational mechanism to improve its ability to respond to highly volatile demand and increasing complexity, to maintain competitiveness by

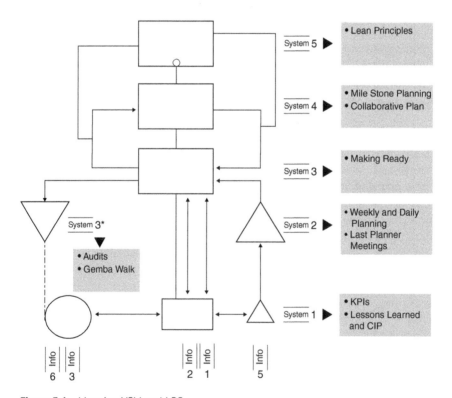

Figure 3.1 Mapping VSM and LPS.

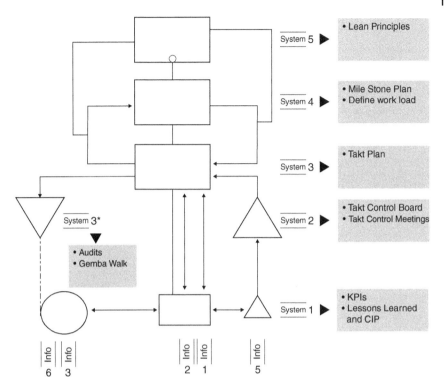

Figure 3.2 Mapping VSM and takt planning and control.

increasing efficiency, to increase customer and employee satisfaction in parallel, and to improve employee retention by increasing task autonomy and consistent employee development.

More detailed information on industry-specific challenges, solutions, and modelling approaches can be found in Section 3.5.4.

Figure 3.3 shows exemplary and in great detail which lean methods are to be introduced and used to achieve maximum organisational effectiveness and economic efficiency. In the authors' opinion, the essential functions here are above all the necessary lean integrations in systems 3, 4, and 5, as well as the connection and emphasis of customer orientation to and in all areas and hierarchical levels of the company.

- **S5 – Lean principles and changed mindset**
  'Tone-from-the-top', the inclusion of complementary topics and standards in the company's value system, the definition of a company's mindset (own reception of a lean production system), cultural change in the direction of

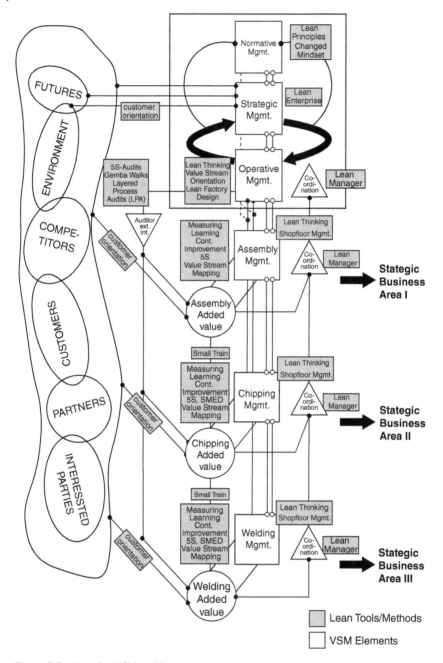

Figure 3.3 Mapping VSM and lean management tools.

self-organisation, lean thinking, integrated management (quality, environment, occupational health, and safety).

- **S4 – Lean enterprise**
  Commitment to the lean enterprise to safeguard the company by increasing efficiency; planning the design of the integrated management system; developing the company and its employees; monitoring new technologies; and assessing consequences for the system, monitoring markets, and competitors.

- **S3 – Resource provisioning and foundations for monitoring**
  Provide financial resources, provide human resources, act responsibly by conserving resources, authorise and organise training, form a task force to create an accepted metrics system for the management system, authorise system 2 (tasks and expectation horizon), target agreement with system 1 management on expected outcomes (budgets, results, further development of the integrated management system).

- **S3* – Compliance with internal and external requirements**
  Monitoring results; ensuring compliance with requirements; determining readiness for delivery; creating necessary documentation; improving processes; increasing efficiency; evaluating training successes; reporting to system 3; monitoring the progress of employee qualification and system development; monitoring the maturity level within the various management system elements; setting up a company-specific integrated management system; conducting internal and layered process audits and accompanying customer, external, and certification audits.

- **S2 – Shopfloor management, coordination, support processes**
  Capacity planning; orientation to customer cycle; levelling of work packages; scheduling of time windows for improvement tasks/maintenance; application of lean methods; pre- and postprocessing of customer requirements and documents; entering into the electronic data processing (EDP) system; timely and economical procurement of raw materials, consumables, and supplies; tracking of delivery dates; request for material certificates; consideration of prohibited substances, conflict minerals; compliance with environmental management and occupational health and safety requirements; provision and maintenance of IT infrastructure, monitoring of PDA; adaptation of the IT system to the needs of everyday business; ensuring usability, data protection, and data security; preparation of evaluations; establishing a training concept; development of rolling and iterative training plans; implementation of training courses; standard reporting; material and certificate management; and internal transport processes; picking and material provisioning; shipment preparation and execution; and customs clearance.

- **S1 – Measuring, learning, and continuous improvement**
  Preparation and processing of customer requirements and documents; project management; make-or-buy decision; updating in the EDP system; application of lean methods; order follow-up; lessons learned, interface and communication function to other systems; preparation of product documentation; initiation and review of machine maintenance; execution of machine and employee requirement planning; updating in the EDP system, application of Lean methods, participation in CIP; participation in training courses; commitment and motivation, compliance with environmental management and occupational health and safety requirements; postcalculation; promotion of cooperative behaviour; identification with the company values; adherence to behavioural guidelines; result orientation; worker self-monitoring; fulfilment of documentation obligations; self-organisation.

3.4 Performance Measurement

3.4.1 General Measurement

In the past, key figure systems were assigned to controlling primarily in finance. In the meantime, however, they have become standard in production or project management. The enormous competitive pressure, the incredible dynamism, or the high level of public attention mean that you have to know at all times where you stand and where you're going. Key figure systems and reporting take up a considerable part of the working time in project management. The project participants have the right to be well informed. But key figures and measurements are also suitable for decision-making, mainly when scenario analyses are carried out. Figure 3.4 shows various general key figure categories.

Key figure systems such as the balanced scorecard (BSC)[4] or the performance pyramid[5] are traditional methods aligned with the corporate vision and strategy and transferred top-down within the company.

---

4 The BSC is used to measure, document, and manage the activities of an organisation in relation to its vision and its strategy. It is a management instrument that uses key figures to measure potential, but in many companies it is only used as a (further) system of key statistics. In addition to the purely financial and result-oriented parameters, nonmonetary elements are also included in the assessment of results. The BSC follows the basic idea of Kaplan and Norton (2003): translate strategy into action.

5 The performance pyramid determines key performance indicators (KPIs) that measure performance in terms of quality or customer satisfaction. It follows a classic top-down principle so that the strategy and expected performance are derived from the company's vision.

Figure 3.4 Key figure system.

Patrick Hoverstadt, in his work *The Fractal Organization – Creating Sustainable Organisations with the Viable System Model* (2009), cites the following reasons for criticising performance indicator systems:

- A group makes unrealistic assumptions of people not directly involved in the production and determines the target values.
- The goals are recorded at a point in time X and are then decisive. Market changes, etc., are no longer taken into account.
- Important topics (e.g. intermediate inputs from suppliers) are hidden and not considered, although relevant for production.
- Risks are not taken into account.
- There is a high control effort to maintain and monitor the key figures.
- Ninety percent of the strategies are not implemented consistently or are not functional (e.g. due to an imprecise planning basis or no person driving the

implementation of the strategy) so that the key figures derived from them are not relevant.
- Results are not used to work together on improvement measures. On the contrary, they mainly serve to assign blame to the other.

3.4.2 Lean Measurement Construction

Performance measurements play a central role in the application of lean construction. Depending on the requirements of the project, it is advisable to develop individual sets of key figures. Examples of lean construction metrics are:

- Process stability (PPC).
- Quantity (target/actual status).
- Productivity (person-hours consumed/person-hours budgeted).
- Number of defects on the construction site.
- Machine utilisation.

The defined key figures and their development are to be visualised with further relevant project information via a control panel on site. The development of the project becomes transparent for all participants. Lean construction key figures show a snapshot of the actual status of the construction site. Actions can be derived if there is a difference between the actual and target state.

Depending on the level of the subsystem, individual key figure sets can be configured so that each team only records those key figures that are helpful. Speaking in VSM terminology, this means that these key figures can, in turn, be aggregated by system 2 and used as a basis for decision-making by system 3.

Key figures should be regularly questioned (e.g. depending on the project phase) and adjusted.

3.4.3 Beers' Triple

In connection with the VSM, Stafford Beer developed the dynamic performance measurement, which is based on three key figures. These are colloquially called Beers' Triple and are reproduced here:

- **Operational performance:** This describes the operational efficiency and the relationship between the provision of resources from system 3 and services from system 1. If there is a gap between system 3 and system 1, this is called the 'efficiency gap'.
- Question: What is currently being achieved with existing resources, and what are we now capable of?
- **Latent performance:** This measures the latent efficiency between system 3 and system 4. Naturally, there is some competition between system 3 and

system 4 regarding resource allocation. System 3 concentrates its resources on the present business, whereas system 4 invests in the future business. A resulting gap is called a 'strategy gap'.

- Question: What are we currently capable of, and what should we be capable of in the future?
- **Organisational performance:** Organisational performance is the ratio between operating and deferred performance. It can be used to monitor the extent to which the strategy has been integrated into the operational units of the organisation and whether the organisation is capable of closing the strategy gap. The strategy gap is closed if the company's strategic needs can be translated and applied to the operative business.
- Question: To what extent is the strategy integrated into the operational areas of the organisational structure?
- Consult Figure 3.5 to help you through the following example.
  - o **Operational performance:** What is currently being achieved with available resources (5 buildings), and what are we presently capable of (7 buildings)? The operating performance is 71%.
  - o **Latent performance:** What are we currently capable of (7 buildings), and what should we be capable of in the future (10 buildings)? The latent performance is 70%.
  - o **Total performance:** To what extent is the strategy integrated into the operational areas of the organisational structure? (5 buildings) are currently achieved to survive in the market, but this should be (10 buildings). The total performance is 50%. The strategy is not integrated.

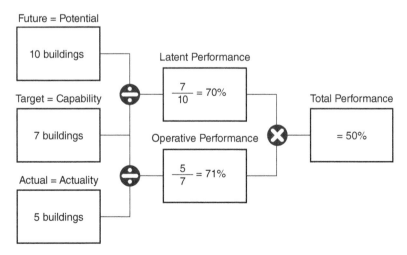

Figure 3.5 Beers' Triple: example buildings produced.

3.5 Case Studies and Practice Insights

To illustrate the preceding theoretical considerations on the VSM and lean tools better so as to present them more understandably, we now present three case studies and a practical application example.

The three case studies and the practical application example are real projects, whereby the case studies are sorted in their size: starting with a small project, through a large project to a megaproject. In the practical application example, a medium-sized company from the mechanical and plant engineering sector serves as a live laboratory.

3.5.1 Case Study: Planning Project

The following is a use case of the organisation of a medium-sized planning project to design a new bridge construction with the perspective of an engineering office based on the VSM.

- **Project boundary conditions:** An engineering office is assigned to prepare the basic evaluation, preliminary planning (variant analysis), and design planning.
- **Project size:** €11.5m.
- **Project duration:** one year, basic evaluation and preliminary planning for six months, design planning for six months.

After placing the order, the engineering office established a project team within the company's bridge construction department, consisting of a drafter, a planner, and a project leader. In addition to planning, the project leader is also responsible for the contractual and coordinating tasks and is the first contact person for the client.

To control the schedule, the project team has set up a planning schedule in coordination with the client, which shows the planning process with its intermediate dates and milestones based on approximately 30 procedures that the project team and the project leader use to monitor progress. Besides, the project leader is responsible for coordinating the planning meetings with the client and the involved trades and persons concerned. At the same time, the project leader must keep an eye on the use of resources and performance and make sure that the team has enough resources to achieve the required performance and results. To assess the relationship between utilisation and deployment, the project leader maintains open communication with their team and involves the owner for a different view. The resource/performance ratio is correct if the project is in place. In addition to the requirements of the customer, which have already changed several times in detail within the framework of an iterative planning process, the planning team must also keep an eye on the requirements of other participants and the requirements from regulations that are subject to specific change cycles.

The project leader responsible for the effective and efficient processing of the project has a good grip on the project with the project organisation they have set up (see Figure 3.6).

The owner of the engineering office, who is superior to the project leader, has all bridge projects of the engineering office under themselves, which are currently seven pieces in new and rebuilt different sizes. As the project leader, they keep everything in view (on the next higher recursion level). In particular, the owner is also interested in the technological development of their industry. They are very interested in the strategic development of their office in building information modelling (BIM) matters. For this reason, they are also a member of a BIM cluster. Based on their market observations, the engineering office had already started five years earlier to plan 50% of the projects three-dimensionally. This helped them to build up the necessary competence. The engineering office is increasingly registering requests from clients to use BIM services. The owner regularly

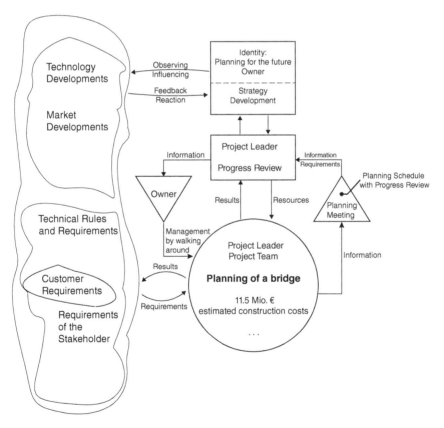

Figure 3.6 Case study: planning project.

involves their project leaders and planning teams in the developments and findings and encourages them to try out new technologies to a certain extent. True to the owner's motto when they founded the office 15 years ago and which still characterises the identity of the engineering office today: 'Planning for the future.'

3.5.2 Case Study: Major Project (Planning and Execution)

This section deals with a field report describing a case study on the application of lean construction, particularly the LPS and management cybernetics (see Figure 3.7).

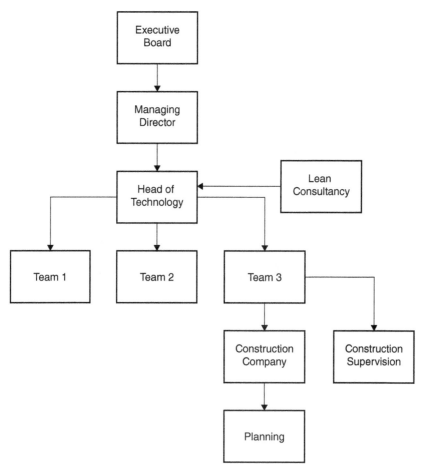

Figure 3.7 Case study: organigram.

- **Project boundary conditions:** Large infrastructure project; approximately 50 participants from the organisational area; approximately 150 industrial employees from production.
- **Project scope:** Approximately €350m.
- **Project duration:** Design phase one year; approval phase four years, tender and award two years, construction phase with construction planning approximately five years.

3.5.2.1 Design Phase and Approval Phase

It all started with the former project manager, a lean construction enthusiast from the very beginning. They brought this enthusiasm with them and thus ensured an essential support element. The operational implementation was carried out by a consulting firm from the lean construction sector.

During the design phase, all participants were trained in the basics of lean construction management during kick-off. The participants were exclusively designers and technical consultants from various specialist areas familiar with production design and planning. They regularly implemented it top-down with a planning schedule based on the specified time schedule. However, to do this using a collaborative production plan based on the LPS was new for everyone involved. This approach was partly well received, and occasionally some old hands rejected the approach.

Establishing the overall process bottom-up, the collaborative implementation with the collaborative plan, the division of the work packages from the different areas, and the establishment of a production plan were good to start the implementation of lean construction.

Also, the measurement using PPC with the measured focal points of the non-compliance of commitments, such as

- lack of advanced performance
- missing quality
- ambiguity

was accepted.

After initial vigour, however, the disciplined handling of the game eased. This was partly due to the local separation of the participants – there was no 'big room' – and partially because the last planner meetings did not take place regularly and sometimes only after longer intervals. One reason for this was the parallel processing, the approval planning, and the objection processing for the plan approval procedure, which was not explicitly 'processed lean'. Thus, the same parties always had to switch their mindset. On the other hand, there was no regular and disciplined work according to lean methods so that a routine could have been established. The support of the external consultants was available.

This support should have been called up more frequently and used instead of the parallel project controller.

Positive

- The support of the project manager was there and therefore critical; it was also a role model and budgeted for.
- Kick-off and competence building of the planning team.
- Establishing a planning schedule according to production planning criteria.

Negative

- The project did not have a 'big room', i.e. no lean environment.
- The participants did not meet weekly.
- The LPS was not performed in a disciplined manner.

3.5.2.2 Tendering and Awarding Phase

The lean implementation strategy was defined during the tendering phase. This included the decision to apply lean construction during execution. Even though there was no support from senior management, there was no resistance.

For the implementation, a separate area, 'lean construction', with its contract position to be priced as a lump-sum position, was worked out within the construction contract. This included all essential basics for lean construction, LPS, integrated project delivery (IPD) team, big room, and cooperative contract obligations like 'talk before writing'.

For the award of the contract, requirements were placed on the future contractor, including on the one hand a letter of intent, which each competitor had to sign, but also the evaluation criteria which had to be taken into account for the award. The internal coordination with the process guardians of the purchasing department and the awarding lawyers was labour-intensive. Both parties had a different perspective – but the agreement was nevertheless successful.

Within the award framework, the subject of lean was a soft evaluation criterion compared to the price. Some bidders did nothing or only the barest minimum requirement (e.g. enclosing their quality manual), while others submitted a bid that included the systematic lean implementation of the construction contract. In the end, the contract was awarded to the bidder with an exemplary lean implementation and an economic bid.

Positive

- Definition of performance in construction contract lays essential foundation stone for implementation.
- A compact but specific definition coupled with a detailed flat rate or exact single position is recommended.

Negative

- Lack of support from senior management.
- Competition only takes up the topic if it brings a 'hard' competitive advantage.

3.5.2.3 Construction Phase

Shortly after the start of the project, a two-day kick-off session was held with all operative participants of the project. The lean construction method was introduced, and there was time for discussions and getting to know each other away from the project.

On the second day, after an introduction to the LPS, an LPS for a real bottleneck was carried out directly. The construction of a structure on the critical path is essential for overall success. The start with a real, vital process was very good. It was also very positive that it was quickly noticed that the last planners were missing. The general contractor outsourced essential services to subcontractors. However, only the subcontractors knew their equipment trains' exact daily outputs and interfaces. The appointment was repeated more effectively and regularly with the subcontractor site managers and supervisors a few days later. The building was completed more quickly thanks to improved coordination and transparency.

In the further course of the project, lean was implemented at various levels of the building implementation. At the strategic level of middle management, regular walkarounds were accompanied by a lean coach as a normative element within a small group where conventions were kept informal, and protocol minimal. After some time, however, these failed because of so-called hidden agendas.

At the operational level, however, a good work culture had developed in the meantime. It was possible to design and build excellent technical solutions together and with trust. A disciplined production planning for the design was introduced, which was scheduled by the work preparation of the building site. There was little waste through contractual correspondence and much-added value for the product.

As it is not uncommon in project business, personnel changes and absences may occur during the project. This was also the case in the present study. Therefore it happened that a young project manager suddenly had to take responsibility for a huge management task with a lot of personnel and a large budget. Unprepared, they took the VSM they were familiar with and designed the organisation based on it (see Figure 3.8). On the one hand, this provided security for an intelligent organisational structure. On the other hand, two critical areas of the organisation were considered in detail: the coordination of the individual teams as a meaningful whole and the communication of the overall structure. After analysing the organisational structure based on the VSM and modelling the communication structure (the meeting arrangements – see Figure 3.9) with networked thinking and system

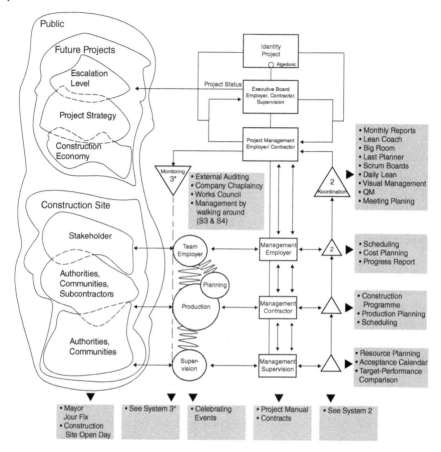

Figure 3.8 Case study: viable system model.

dynamics methods, the organisation was built. As a result, the positive boundary conditions saved half a year of construction time.

3.5.3 Case Study: Megaproject (Execution)[6]

3.5.3.1 Boundary Conditions

The project is a megaproject with a budget of €4 billion. The lengthy preparation period (including awarding of contract) was 20 years and has now been completed. The construction phase has begun. The megaproject is being financed by a consortium of private and public investors. Owing to the high project volume and the

---

6 Based on Frahm and Rahebi (2021).

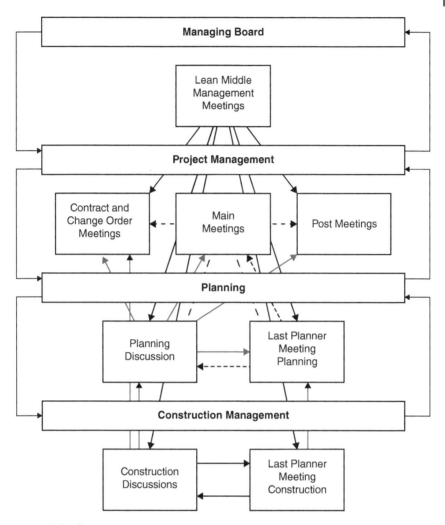

Figure 3.9   Case study: system dynamics model meetings.

associated risks, the client decided during the planning phase to establish its own project organisation (PO).

As can be seen in Figure 3.10, the project company is managed by three managing directors (the chief executive office, the chief technical officer, and the chief financial officer). The overall project is divided into several sections. Each section is represented by a technical and a commercial manager with responsibility for the results. In addition to the employees in the project teams of the sections, there are staff functions for coordination at the programme level.

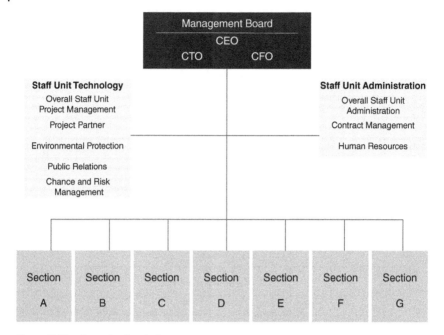

Figure 3.10 Organisational chart.

Figure 3.11 shows how the project organisation fits into the parent company and illustrates the principle of recursion levels and the four recursion levels of the case study organisation.

As with many large companies, the parent company/group (recursion level −2) consists of several thousand employees and owns innumerable viable systems and many subsystems and producers. As an example, Figure 3.12 shows the infrastructure division, which carries out infrastructure projects, and the passenger transportation division, which is responsible for transportation services. Both divisions are, by their very nature, interdependent.

With Figure 3.13, a closer look at the infrastructure division (recursion level −1) is possible. Operationally managed by the infrastructure board, various activities exist within the infrastructure division. One of these, shown as system 1, is the project organisation of the megaproject.

The strategic environment, ministries, market trends, and the economy all play a decisive role here. The operational environment includes complexity drivers such as local and national politicians, authorities, potential customers, and infrastructure project operators. In contrast to other megaprojects implemented by the infrastructure division, the project organisation has the advantage of being

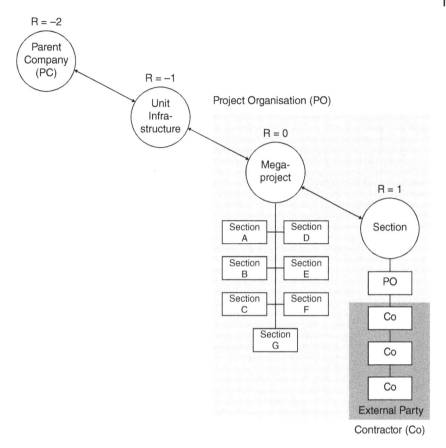

Figure 3.11 The organisation as a whole.

an independent company. This means a certain degree of freedom in operational and strategic decisions at the project organisation level.

Brookes (2015) states, based on her studies, that own project organisations implement 50% of all megaprojects and that this has a positive influence on cost development and the stability of the construction schedule. However, it is essential to note that such an organisation must be set up in good time before the start of the project and adjusted according to its progress. Thus, the personnel expenditure during the planning phase is lower than during the construction phase. Greiman (2013) concludes that a separate project organisation is suitable for the efficient use of resources and recommends an advisory board to accompany it.

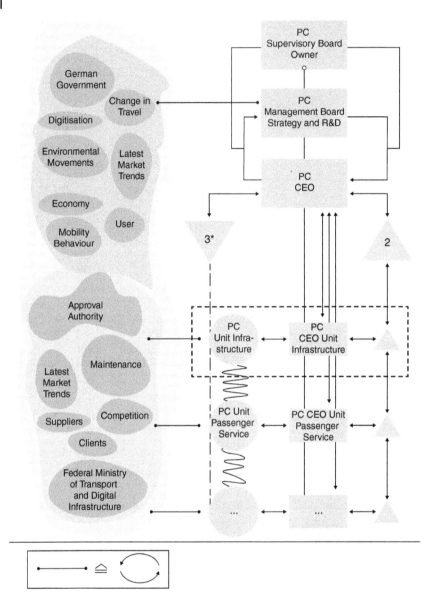

Figure 3.12 Level −2, perspective parent company/group.

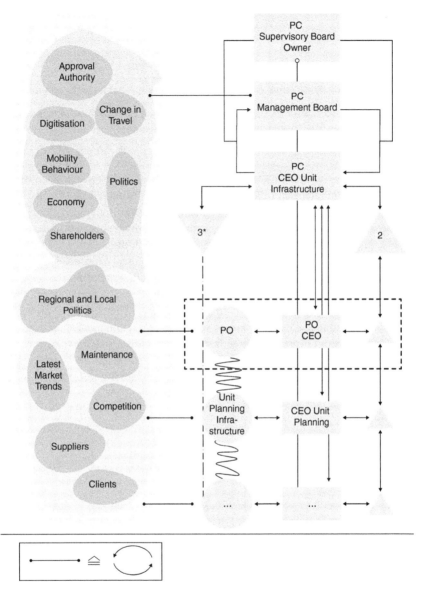

Figure 3.13   Level −1, perspective infrastructure division.

3.5.3.2 Analysis of the Megaproject

The further consideration of the organisation is done in two steps:

1. The consideration of the overall project from the perspective of the programme level.
2. The specific organisational consideration of Section A of that megaproject.

Figure 3.14 (recursion level 0) shows the overall structure of the megaproject. It is derived from the higher recursion levels; the supervisory board, the owners, and the chairperson of the management board act as a normative element (system 5) embodying the values and guidelines of the parent company.

System 4, which deals with the strategic development of the megaproject, is represented by a three-tier team. The advisory board consisting of experts serving as a medium for best practice and enabling practical solutions about the long-term perspective. From a strategic point of view, the infrastructure board represents the parent company's interests and considers the overarching objectives. The management board is responsible for implementing the necessary strategies and activities from the perspective of the megaproject and for incorporating the required information in operational management and production. The operative management (system 1) consists of three managing directors. They are responsible for the provision and use of resources and the operative implementation and performance of the overall project. Section managers are responsible for their sections and can independently carry out day-to-day business and operational decisions up to certain thresholds. Cyclical reporting is used to measure the performance of the sections.

In addition, there are various coordination units (system 2) at the programme level for overall control. This enables the management to engage with overarching concerns, and to adapt issues and findings that affect more than one section. The cycle is closed by external audits (system 3*) at the programme level.

3.5.3.3 Section Analysis

Figure 3.15 – Level 1, perspective section shows the last recursion level, recursion level 1, with the subsystems at the project section (in size a megaproject by its own) level of Section A.

Section A's project management establishes section-specific standards within the framework of its autonomy, which is reflected in the coordination and monitoring of the subprojects. In addition to the overall coordination of the staff units, Section A aims to establish its own culture, which promotes partnership with contractors and external stakeholders. Workflow management instruments such as lean practices are used for situationally appropriate coordination. Monitoring includes methods such as 'management by walking around' or regular 'coffee and issues meetings' for the exchange and sporadic collection and verification of information.

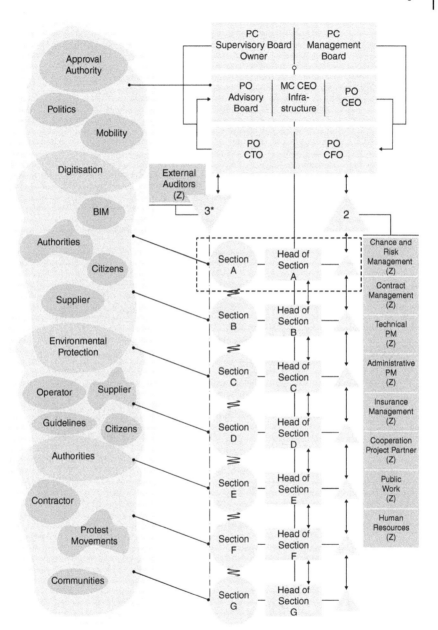

Figure 3.14 Level 0, perspective megaproject.

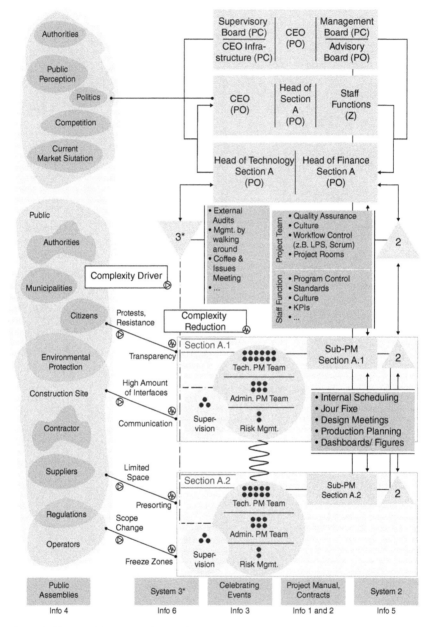

Figure 3.15 Level 1, perspective section.

The subprojects are similar in structure, and the subproject managers each lead a team consisting of technical or commercial project managers, risk managers, and construction supervision. The two subproject managers are in close contact and thus benefit from synergy effects. The subproject-specific coordination methods are identical for both sections.

The primary complexity drivers of the environment consist of the categories public (e.g. authorities, citizens), building components (e.g. main contractors, suppliers), and operators. The analysis of the information channels shows that project manual and contracts are the main media for internal communication. On the environmental side, events that explicitly inform stakeholders about the project and involve them in the project have proven to be positive.

3.5.4 Practice Insights from a Medium-sized Mechanical Engineering Company

3.5.4.1 Challenges for the Industry

With 1.029 million employees (related to companies with more than 50 employees), the German mechanical and plant engineering sector will remain the largest industrial employer in 2020 – despite a weak economy due to the pandemic, an uncertain outlook for the future, technological change in general and structural change in the critical sector of vehicle manufacturing in particular, as well as the potential threat of liquidity bottlenecks in 2021 (Paul 2020).

Different other risks jeopardise positive corporate development. In addition to the factors already mentioned, the Chamber of Industry and Commerce of Lower Bavaria identified the local shortage of skilled workers, relatively high labour costs, as well as energy and raw material prices in its economic report for the beginning of 2021 (Industrie- und Handelskammer für Niederbayern 2021).

To be able to be successfully active in the market and exploit the potential of the industry despite these problems, there is a need to 'find the right answers to the challenges', according to Karl Haeusgen, the President of VDMA (German Mechanical and Plant Engineering Association) at the association's virtual annual press conference (Paul 2020).

Secondary to the general problems listed here, which apply to the industry as well as the respective location of the company, each company also faces individual challenges, which can result from a variety of factors and their interactions. For JELBA Werkzeug & Maschinenbau GmbH & Co. KG, headquartered in the extreme south-east of Germany, these include, on the one hand, the broadly diversified product range, which extends from medium-volume machined parts, through precision machined parts with a unit weight of up to 80 t, to complete systems including the assembly of all the necessary electrical, pneumatic, hydraulic, or hydrostatic components. On the other hand, despite its location in a market niche, the company serves around 10 industrial sectors, which are characterised by the strongly differing customer and standard requirements and accompanying regulations (JELBA Werkzeug & Maschinenbau GmbH & Co. KG 2019/2020).

Over and above these requirements of external stakeholders, there are also often implicit demands on the company, most of which stem from the workforce and its environment, such as occupational health and safety, ergonomics, training of young skilled workers, guaranteeing a safe and attractive workplace, and opportunities for personal and professional development. The synopsis of these diverse requirements and objectives results in a wide-ranging field of tension that illustrates the rapidly increasing complexity to which production companies are exposed in their everyday business, as also exemplified by the Complexity Management Academy of RWTH Aachen in its documents on the topic of complexity management (Complexity Management Academy – Abteilung Innovationsmanagement am WZL der RWTH Aachen 2020).

Furthermore, there are necessary activities to ensure the future viability of the company and the further development of existing business models based on current market trends that are already emerging. The 'German Engineering' study names medium- and long-term trends in this regard, among others: application of new disciplines (e.g. 3D printing), expansion of the range of services, consideration of current consumer trends (digitalization, batch size 1), engagement with artificial intelligence and advanced analytics, as well as orientation towards new sales markets (FTI-Andersch Consulting 2021).

Together, these factors form the challenge that small and medium-sized contract manufacturers must face now and in the future. Therefore, they also serve as a starting point for the initiated change processes, examined in more detail throughout this chapter. Both the exact requirements and challenges with which the company in this example is confronted as well as the desired target state are aggregated individually for each company in Figure 3.16.

3.5.4.2 The Solution: The Creation of a Hybrid Corporate Form Based on the VSM

To consider a large part of the company's internal objectives and external requirements, in particular customer wishes, when creating a suitable, sustainable corporate structure, specific emphasis was placed on combining effectiveness and efficiency within the system to be made. This approach should combine organisational, process-related, and economic advantages so that a win–win situation is designed for the company (and its employees) and its customers.

This is to be achieved by using the viable system model (VSM) as a framework structure (see Figure 3.17) and by integrating methods, techniques, and tools from the disciplines of lean management and quality management, whereby the great

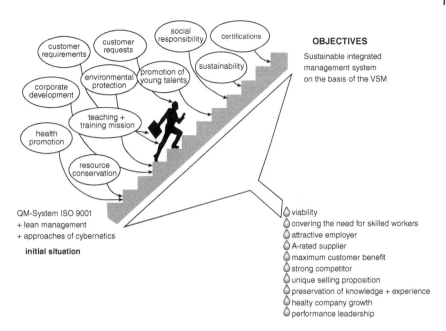

Figure 3.16 Tension between internal and external demands and objectives in medium-sized production companies.

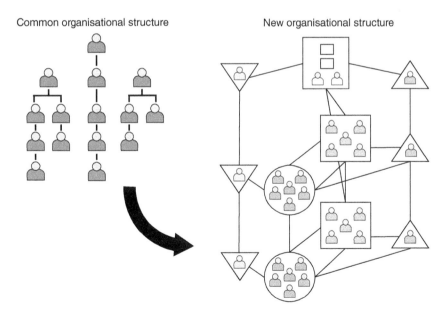

Figure 3.17 Transformation of a hierarchical organisation structure into a viable system.

challenge is to combine all the laws, specifications, and standards on the customer side with the striving for task autonomy and the possibility of free professional development on the employee side as harmoniously as possible. At the same time, it should be ensured that the company can retain and expand maximum flexibility in dealing with the constantly changing requirements.

Owing to its universal applicability to all kinds of natural life forms (plants, animals, humans) and abstract organisational forms (companies, states), the VSM offers the perfect basis, describing an interaction with their environment. This gives the model its special responsiveness to changing framework and environmental conditions, allowing it to initiate appropriate reactions immediately. In the language of the organisation, this represents the anchoring of strategic planning.

In addition to strategic planning, the model has a superordinate instance, system 5, in which values, consciousness, or ethos are anchored. Through the self-similarity of the system, the validity of all these values and laws can be derived for all subsystems and their elements. Unfortunately, the generally binding nature of the corporate regulatory framework often degenerates into a purely formal commitment in various other forms of organisation.

The VSM also has an additional control instance that additionally verifies the results reported by the operational systems to the management systems. The (cybernetic) control loop model corresponds to the comparison of target and actual values (Wiener 2013) representing the audit channel in the operational context. This institution is particularly interesting for companies located in the area of contract manufacturing, as they are obligated to particularly extensive verification, validation, and the associated documentation measures vis-à-vis their customers.

Finally, the use of closed information and communication loops has the invaluable advantage of ensuring that information is passed on to its destination. Particularly in the operational context, a not inconsiderable proportion of errors and deviations arise from the loss of information and interface problems. Ensuring that information is passed on through closed communication loops in the operational organisation is essential for functioning processes and eliminating interface problems.

3.5.4.3 From Theory to Practice: The Organisational Structure

After these general considerations, a great deal of (time) effort was invested in the definition of system 1s, i.e. the company's strategic business units. On the one hand, ensuring viability was the subject of intensive consideration; on the other hand, strategic considerations regarding the future development of the business units were also necessary for a final definition of the system 1s.

In addition, owing to the company's product portfolio and departmental structure, the challenge arose that the identified system 1s had to be viable (economically) on their own and function as internal customers or suppliers as needed. In a further step, the basic structure of the organisation was thus created with four system 1s, whereby the four systems can supply each other from bottom to top along the value chain in Figure 3.18. It was also decided to deliberately include the 'amoeba' on the left-hand side of the model in the figure.

The reasons for this were the emphasis on customer and service orientation, which is visible to all employees, partners, and customers, as well as the permanent strategic review and, if necessary, strategic (re)direction of the company to the requirements of the environment.

The newly created organisational structure with four system 1s based on the VSM can be seen in Figure 3.18.

3.5.4.4 Levels of Complexity

The manifold demands of the environment on a company include, above all, the continuous development of skills and capabilities (of machines, plants, but also of people and processes) with a simultaneously economic, even competitive, pricing. To define a rational basis for assessment that allows the allocation of resources in line with expenditure, but also to show the current status with development potential in this area, a gradation into three levels of complexity was also developed based on a total of 10 criteria. The criteria to be considered include costs, deadlines, risk, precision level, measurement, required qualifications, project management, documentation, handling, and certifications. These are differentiated into three levels of complexity (levels 1–3) that build on each other (see Figure 3.19).

The assignment to the respective subcomplexity level results from the values achieved in the calculation or evaluation of the assigned key figures. These are determined during the preparation of the offer based on empirical values and by scaling project data of related orders, or derived directly from the customer's wishes or requirements (see Table 3.1). The mean value formation results in an overall complexity level for an order, whereby individual criteria have a higher weighting. Particularly critical or relevant criteria can override the formation of the mean value if this is considered necessary or is required by the customer (Figure 3.20).

Depending on the respective level of complexity, the number of process steps required increases up to the maximum level of the process.

In addition to economic considerations, the focus here is mainly on the systematic development of the company's ability to successfully deal with complexity and dynamics – despite, or precisely because of, the strongly differing requirements depending on the industry supplied. In addition, the implementation of such a categorisation of complexity levels ensures a certain degree of flexibility in resource allocation for the company.

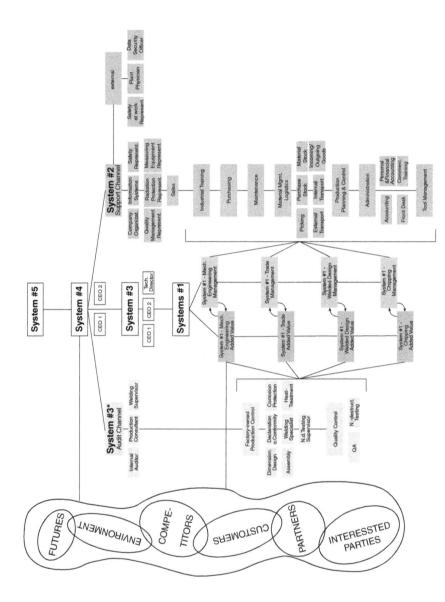

Figure 3.18 Newly created organisational structure with four system 1s based on the VSM.

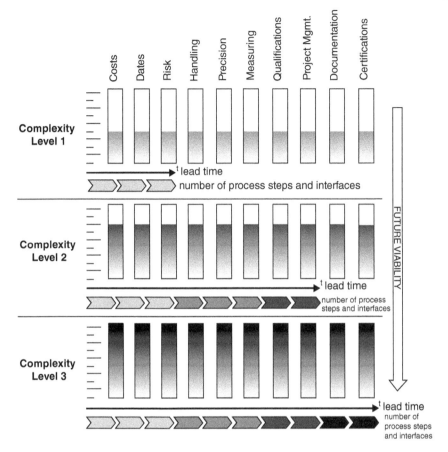

Figure 3.19 The ten criteria for defining complexity levels and their correlation with the number of required process steps.

3.5.4.5 Process Organisation

After the definition of elements that are more akin to the classic organisational structure, the corresponding counterpart in the area of the process organisation has also been created (Pfeifer and Schmitt 2014). The existence of an organisational chart and process map is important because these representations are traditionally frequently used in the industry under consideration. In addition, the company's customers regularly request these representations as part of supplier approval processes and customer audits.

Through the intensive design work on the newly created organisational chart, it became apparent that by tilting the representation of the VSM organisational chart 90° to the left the basic structure of the process map, including the rough process

Table 3.1 Complexity criteria with assigned key figures and gradations.

Criterion	Characteristic/Indicator	Complexity Level 1	Complexity Level 2	Complexity Level 3
Costs	Total project turnover	€1–50 000	≥€50 000/100 000	≥€1 000 000
	Total project costs	1– €	≥ €	≥€
	Requirements for materials	Standard material possible exclusion of individual countries of origin	Standard material, batch material	Standard material, batch material, special materials,
Dates	Total project duration	Standard project duration		>5 months
	Deadline	Buffer time ≥5–6 weeks	Buffer time <5 weeks	Time-critical at project start
	Number of internal interfaces	≤10	10–19	>20
	Number of external interfaces (service providers)	None	1–2 external standard procedures	>2 external standard procedures, time-critical external procedures, new external service provider
Risk	Part type	Noncritical part	Critical part	Safety-related part
	Production frequency	Repeat part	New part	New part
	Occurrence of production errors	No/uncritical previous errors	Previous errors	Previous critical error event
	Project clarity	Technical clarity	Technical questions open	Technical uncertainties
	Risk of accidents	Low	Medium	High
Handling effort	Weight	0–5 tonnes	5–20 tonnes	>20–80 tonnes
	Dimensions	<1000 × 3000 × 10 000 mm	>1000 × 3000 × 10 000 mm	Maximum dimensions
	Restrictions on equipment	None	None	Customised coolants, separately approved component marking
	Special requirements for	None	Defined project area	Clean room

Transport requirements/restrictions		None	Special requirements regarding loading necessary, loading restrictions, forward goods, express goods	Specifications regarding the forwarding agent to be commissioned, heavy transport, transport of dangerous goods, express goods, air freight
Required level of precision	Defect rate			≤1.5%
Internal deviations		Not to be expected	To be expected, reworking on own initiative	Expectable, notification to the customer
Industry		Mechanical engineering	Mechanical engineering, automotive industry, food industry, aerospace technology	Mechanical engineering, automotive industry, food industry, aerospace technology, nuclear technology
Process complexity		Simple	Medium	High
Measuring effort/accuracy	Total number of tests required	Simple final check		Initial piece inspection, various intermediate inspections according to customer specifications (measurement plan), specification of the measurement procedure
Required measuring equipment		Hand-held measuring equipment	Measuring machine	Measuring machine + protocol
Tolerance ranges for technical		In the range of 1 mm		<10 mµ
Requirements for calibration of measuring equipment		Regular calibration of the measuring equipment	By certified test laboratory	By accredited test laboratory
Type of tests		Visual inspections, simple dimensional checks	Visual inspections, dimensional inspections, external tests	External tests, material laboratory, UV crack testing, initial sampling, obligatory presence of an external inspector/customer

(Continued)

Table 3.1 (Continued)

Criterion	Characteristic/Indicator	Complexity Level 1	Complexity Level 2	Complexity Level 3
Required qualification level	Additional training per employee	None	None	Safety culture, nuclear safety culture
	Special qualifications required	None	Crane licence, forklift licence	SCC training, radiation passport, safety check, special personnel certifications
	Occupational health examinations	None	Proof of visual ability for visual examinations	According to StrlSchV, welding medical examination
Project management effort	Occupational health examinations	Complete documents and drawings without special requirements, possibly existence of 3D models		Use of checklists FB-…, compliance with customer-specific standards and other specifications
	Number of internal process	2	3	>4
Documentation effort	Type of documents	Standard documentation	Material certificates	Preliminary examination documents,
	External review of	None	By customer	By (accredited) certifier
	Scope of documentation	No separate documentation	Up to 50 pages of documentation	>50 pages of documentation
Desired certifications	Quality management (DIN EN ISO	Mandatory	Mandatory	Mandatory
	Environmental management (ISO	None	Desired	Desired to required
	Occupational health and safety management (ISO 45001)	None	Desired	Desired to required
	Special standards	None	Factory standards	KTA 1401, ASME, SCC

StrlSchV: an abbreviation for the German word for 'radiation protection ordinance'.

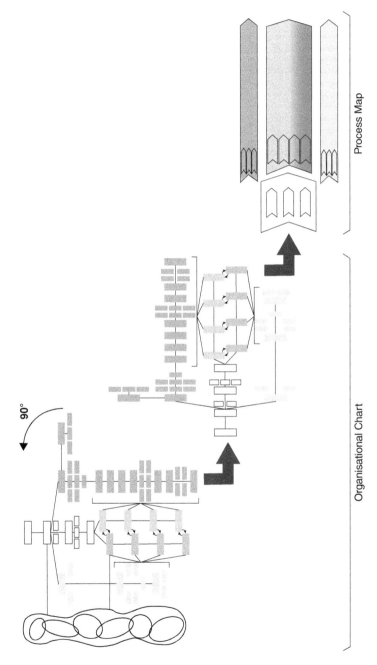

90°

Organisational Chart

Process Map

Figure 3.20 Derivation of the process map from the VSM organisational structure.

descriptions, can be derived relatively easily (see Figure 3.20). This close relationship in content, which is extremely rare in traditional hierarchical organisational charts and the corresponding process maps, is one of the significant advantages of the method described. It makes the connection between administrative units and process flows clearly recognisable for the employees and other viewers.

3.5.4.6 Role Profiles

To deepen this transparency even further, role profiles have been introduced instead of the previously used job descriptions, the categorisation of which can be derived directly from the process types of the process map. Thus, each employee no longer has a single job description but a role profile, which can consist of several different roles, whose content orientation and location in the organisation can range from correlating to indifferent to competing (Merton 1995). The newly created role descriptions contain a set of values and norms derived from the management processes and core and support tasks from the associated main and support processes. In addition, control tasks derived from the audit channel of the VSM system have been added to the method (see Figure 3.21). Since control and inspection tasks have always been part of the company's services as a service provider for machining and assembly orders, it was entirely appropriate to anchor them as a separate process strand in the organisational structure and process organisation, thus emphasising their importance even more prominently.

To be able to apply the full effect of the new method, after the redesign of the structural and process organisation, a corresponding setup for the use of the VSM as a communication and control tool is still missing, compare "Die dritte Dimension des Organisierens" (Pfiffner 2020), which is described in Section 3.7.4.

3.5.4.7 Organiplastic as a Base for the Management Cockpit

Owing to the high complexity of both the internal and the external processes of the company, the entire system can only be mapped with the help of software support and ultimately used as a control system. In order to demonstrate the systematic structure of the system, at least in one area, a 3D model was created by hand using the simplest of means: so-called organiplastic (Figure 3.22).

The advantage of organiplastic over 2D representation is the approximation to the structural complexity of the model as well as the possibility of representing the different levels of recursion and complexity in the area of value-creating systems.

In a further step, this 3D representation was subjected to digital modelling, which forms the basis for the additional work steps (Figure 3.23).

As part of the further development of the model, the IT systems used internally (planning tool, enterprise resource planning, or ERP, system, quality management, or QM, system, production data acquisition) are connected and function as a 3D management cockpit.

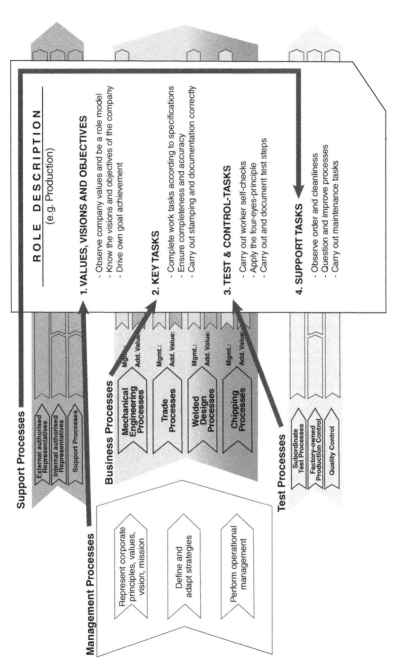

Figure 3.21 Creation of the new role profiles based on the assigned processes of the process map.

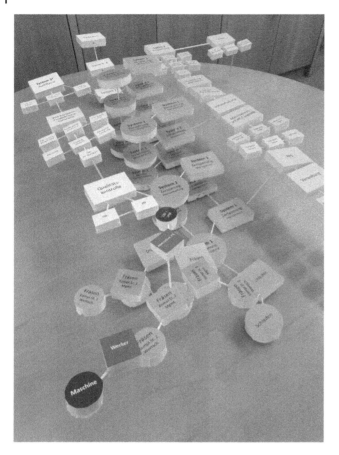

Figure 3.22 The organisational chart of the company as a simplified 3D model (organiplastic).

In addition to classic business management surveys, evaluation, and the installation of controlling and planning instances, the management cockpit also provides the data basis for anticipatory machine maintenance, the digital documentation of production and testing processes, and the resource-appropriate allocation of orders as well as other key figures from the interface area of business management and efficiency increase through lean management (Michalicki and Schneider 2020).

From the perspective of QM, the cockpit can be used to map order-specific organisational charts and process structures for (customer) audits, among other things – for employees to locate their contribution to the overall project, for project managers to permanently check the status in real time, and for the customers to precisely track deadlines and an overview and use of technical parameters of the work done during its creation (see Figures 3.22 and 3.23).

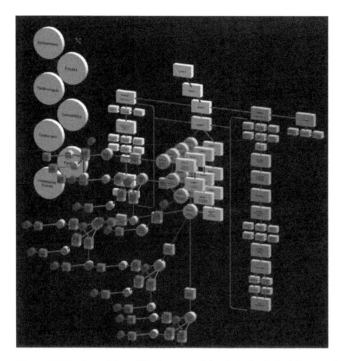

Figure 3.23   From 2D to 3D: the organiplastic in its full digital version.

3.5.4.8   Conclusion

In summary, it can be stated that the newly developed method described (RoBau-Method©) simplifies the handling of complexity and dynamics, which is based in particular on the classification of projects according to complexity levels and the resulting allocation of resources according to the effort.

In addition, the method proactively supports the development, formulation, constant adaptation, and communication of the corporate strategy, which is an indispensable building block for achieving the desired sustainability of a company. The method's most significant potential, however, lies in its ability to unite the most diverse demands of the various external and internal interest groups in the best possible way so that all those involved can participate in the design of the respective processes and thus optimally benefit equally from the success of the company. This is briefly described again here for the main interest groups:

- From the perspective of the company's (potential) customers, the method, first of all, ensures compliance with (factory) standards and laws as well as other desired technical and other requirements when processing an order. In addition, compliance with these requirements is verified and, if necessary, validated in due detail through the development of a separate audit channel.

Customers also benefit from the continuous improvement of processes and simultaneous application of efficiency-optimising measures by obtaining high-quality products and using innovative approaches in technical product design and manufacturing.

Finally, the permanent comparison with and, if necessary, the readjustment of the corporate strategy bring benefits for customers through the expansion of the product and service portfolio by the provider.

- For the company, the method brings advantages both at the meta-level and in the context of business management practice.

The method enables system 4 to work actively on and with the corporate strategy. The binding nature of the goals formulated in this process, together with the values and norms of system 5, is integrated as equally imperative in all system levels through recursiveness.

In its form as a control tool, the VSM opens up a high degree of transparency of all processes for the company using it, which is achieved through the horizontal and vertical communication channels and the results of which are verified through the activities of system 3*.

For this purpose, a separate digital management cockpit is created, which is intended to provide the best possible support for entrepreneurial decisions within the framework of corporate management compare (Hetzler 2008). The work processes achieve previously unavailable transparency and thus an improved basis for production-related and business management decisions.

Furthermore, the method emphasises the management of system 1s and their responsibility, which is already supported by the application of shopfloor management in this company.

About human resource management, the method is interesting for the company in two respects: the cooperation in and between the systems as well as the increased provision of information through the internal networking of the IT systems can ensure the transfer of knowledge between the employees, which, if consistently promoted, leads to the creation of a learning company. In addition, employment in the company becomes more attractive for employees because of the opportunity for participation, professional and personal development, proactive shaping of the company, as well as career planning for the individual (in the highest expansion stage), which in turn has a positive effect on employee retention and length of stay in the company.

- In particular, the employees benefit equally from the advantageousness mentioned above in the area of human resource management: from the employees' perspective, the possibility of participation and a certain degree of task autonomy, in particular, are factors that have both a motivating effect on the individual and lead to an overall positive working atmosphere. An additional employee-oriented factor is the extensive abolition of departmental boundaries and the promotion of interdisciplinary work, which takes place, for example, within the framework of project teams.

The development of personal and professional skills and abilities and the associated opportunity for advancement within the company is also an attractive incentive for employees to become involved in the company, develop themselves, and remain with the company in the long term.

3.5.4.9 Adaptability

Finally, it should now be briefly considered whether and how the applied approach can be adapted for other companies.

Owing to the central objective of the method, which aims to combine external requirements with internal development and design potentials in the best possible way, the approach is probably well suited for companies that are in a similar position in the supply chain as the company under consideration, i.e. suppliers or contract manufacturers without their own product, which therefore have to orientate and align themselves firmly to the demands of the respective customers and at the same time are subject to increased competition and cost pressure. The method can also be used for companies with their own product, but the handling of complexity could focus more on a different aspect.

About the applicability in different sectors, no limitation is to be expected. Still, the method's advantages in dealing with complexity can only be fully exploited when several sectors with more or less strongly differing requirements are supplied.

About the suitable company size and organisational form, it is predicted that the method is well suited for all company sizes from small to medium-sized companies with batch sizes in the range of single-item, variant, and small batch production, since interdisciplinary work is often (still) part of everyday business here. In very large companies with highly fragmented and small-scale organised process flows as well as a highly pronounced process depth, such as in large-scale, mass, or clocked production, it can be assumed that the desired freedom in the context of work design and self-organisation approaches is neither organisationally feasible nor desirable in terms of the process capability to be achieved.

3.6 Findings, Criticism, and Reflective Questions

3.6.1 Findings

Lean should be understood as a systemic approach. It is the systemic, strategic orientation of the organisation towards flow efficiency and cooperation.

According to Iris Tommelein, the management cybernetics and, in particular, the VSM can be used as a reference and analysis model for implementing lean construction (Steinhäusser et al. 2015). This applies to the LPS and the cycle production.

According to Elezi and Steinhäuser, cybernetics is a first-order principle (truths, theoretical, abstract) and lean construction a second-order principle (instructions for execution, practical) (Steinhäusser et al. 2015).

Recommendations for implementation of lean construction

- Daily meetings (many, short, and quick decision-making).
- Involve authorities, approval experts, and technology experts inherently in the design and planning process right from the start.
- Initiative 'speaking before writing' (many people are/would like to be lean/ cooperative).
- Be decision competent (no decision backlog on the sides of the client/ contractor).
- Let execution know-how flow from the contractor into the tender/award phase.
- Streamlining of the planning processes on the part of the client (fewer test instances).
- Ensure the presence of a power promoter.
- Hidden agendas can jeopardise implementation.

Effectiveness and efficiency are represented by the VSM and lean management combination.

- Effectiveness means 'doing the right things' ensured by VSM.
- Efficiency means 'doing things right' ensured by lean management.

In terms of practical implementation, the VSM can be applied as a structuring element, with the primary objective being the system's effectiveness. This ensures that a change in the mind of the company's mindset is achieved during the long-term application of lean and that it is not terminated prematurely without results.

In this connection, the second element, lean management, operationalises the cybernetic system, which is kept very general and is often challenging to grasp. The very operational lean methods implement the VSM more tangibly for employees. In addition, the objective of lean is to increase efficiency, which, in the case of a successful connection, results for the company in a smoothly functioning organisational structure and process organisation and economic advantages.

The collection of information through key figures is essential for the information of participants, well-founded decisions, reporting, and the control of the project.

When selecting key figures, the need must be taken into account. It is advisable to record a few but meaningful key figures, which must be regularly questioned and adjusted.

Lean construction key figures provide process-oriented production control about quality, deadlines, and costs.

Key figures can be visualised at the production site, e.g. using a control panel.

With Beers' Triple, the strategic dimension of the organisation is considered in addition to its operational dimension.

3.6.2   Criticism

Generalisable advice for adapting the examples shown can only be given to a minimal extent since the system architecture must be individually adapted to the respective industry and organisation.

Combining two complex methods, such as lean management and the VSM, increases the demands and challenges for change management during implementation and the further sustainable operation of the system.

The generally relatively rare application of the VSM in practice results mainly from the perception of the model as complicated and/or too abstract.

Lean management is often misinterpreted as a toolbox, whereby the meaningful lean philosophy is lost. This often leads to the failure of lean implementations.

Owing to the manifold and technically very differentiated lean tools, a selection must be made individually for each company. It is essential to consider the product and service portfolio and customer requirements.

When using lean KPIs as a basis for benchmarking and as a basis for decision-making, it must be taken into account that applicability, meaningfulness, and comparability are strongly dependent on the respective industry and company size. If the congruence in these areas differs, the meaningfulness of the KPIs also decreases accordingly.

3.6.3   Critical Reflection to Practice Insights from a Medium-sized Mechanical Engineering Company

Even though the VSM in general, and the method described here in particular, appears to be well suited to function both as an entrepreneurial steering tool and as the basis for a new, less hierarchical organisational structure, there are nevertheless some issues that need to be critically noted.

It can be assumed that the adaptation of the method will not produce similar benefits for all users. For example, large companies and producers in the high-volume unit number range tend to be less suitable, but this would still have to be verified in related research projects.

Furthermore, the method described is a very pragmatic interpretation of VSM, which in certain areas represents a compromise between traditional VSM structures and elements of project/process and lean management and deviates from the original conception of the model.

About the various new forms of work, it should also be pointed out that, despite the striving for self-organisation, a completely agile organisational concept or complete democratisation of administrative and work tasks can probably not be achieved with the method described. However, typical users of the technique will be able to establish a hybrid organisational form, taking into account all external requirements.

3.6.4   Reflective Questions

- What are the benefits of combining lean management and the VSM?
- Is the connection between efficiency and effectiveness equally significant for every organisation?
- What could cause deviations in the importance of efficiency and effectiveness?
- Does this situation represent an opportunity or a risk for implementation?
- Does an organisation's preference for efficiency or effectiveness affect the system's design?
- And if so, what would be the impact?
- How can audit channel 3* be used as both an information and trust channel for the organisation's management without running the risk of employees misinterpreting the audit channel as a monitoring tool?
- In which areas of application are KPIs indispensable? Where are they possibly helpful and where a hindrance?
- What needs to be considered when adapting the approaches from the case studies for your own company?
- Which methods are universally applicable and which are more industry-specific?

4

Beyond Cybernetics and Lean

If you defer investing your time and energy until you see that you need to, chances are it will already be too late.

Clayton M. Christensen

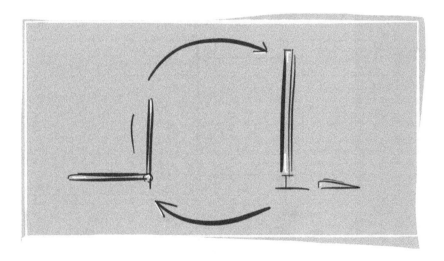

Designing Intelligent Construction Projects, First Edition. Michael Frahm and Carola Roll.

4.1 Control, Regulate, Steer

Construction projects are managed through controlling, regulating, and steering (Figure 4.1). The key characteristic of controlling technical systems in control engineering is that the output variable does not affect itself again through the input variable. There is no feedback. In the project business, the control concept is broader.

The regulation is a matter of processes in systems in which interactions occur, and a variable quantity is kept constant independently. A regulation process is a targeted influencing of variables in systems. Generally known applications of technical systems are, for example, the heating control of the iron or the cruise control of a vehicle. These can be the classic control variables in the construction industry, such as costs, deadlines, and quality management.

Steering is described as the ability of a system to keep itself under control. This means that a system must have a sufficient degree of freedom to act and should only be exposed to those boundary conditions in which its steering can function. See also Ashby's variety theorem and Luhmann's selectivity strategy

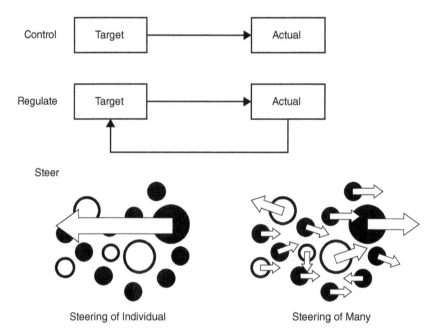

Figure 4.1 Controlling, regulating, and steering. Steering (left graphic): direction and activity are done by action and decision 'top-down'. If the top-down decision-making power fails, the system is incapable of action. Steering (right graphic): direction and activity are performed by the action of many. There is a balance between top-down and bottom-up.

in Chapter 1. Closely connected with the concept of steering is the concept of self-organisation. If a system has a good ability to organise itself, it also has a high steering ability.

Every individual in an organisation should have the greatest possible room to manoeuvre. Using a construction project as an example, this should apply to the entire hierarchy, i.e. the workers, the foreman, the site manager, the senior site manager, etc.

Everyone should and must be able to decide. Suppose everyone acts as a 'manager', i.e. as someone who operates something like an engine within an organisation, within the framework of their abilities and competencies. In that case, the structure approaches a higher degree of self-organisation and thus of control. There is no need to draw a clear line between the system and the driver. Steering must be part of the system. Steering is not by a person but by the behaviour of the system.

4.2   Self-organisation

One evening, two CEOs sit together at the fireplace of an up-market hotel and talk things over. One of the CEOs has to interrupt the conversation all the time because his mobile phone keeps ringing as one of his employees keeps asking him for decisions.

This is visibly unpleasant for the CEO, and he apologises to his counterpart for the rudeness. But he notices that the mobile phone of the other CEO does not ring at all!

So he asks her:

> 'Say, why doesn't your phone ring?'

The other CEO responds with a smile:

> 'We have now organised our company in such a way that the employees can make their own decisions!'

What is this short story supposed to tell us? Self-organisation makes the lives of board members and superiors easier and strengthens the ability to organise oneself.

Even before the beginning of the digital age, Drucker described the rise of 'knowledge workers' who would increasingly use their intellect as their power to create value (Drucker 1969). At that time, Drucker already called on managers to bring decision-making and responsibility as deep (bottom) as possible into the organisation. Successful organisations in the digital age do just that!

However, to be able to implement this successfully, the following prerequisites must be taken into account.

- By the congruence principle of the organisation, the unity of task, competence, and responsibility must be urgently maintained in the delegation (Reiß 1982). This means that, when delegating tasks, not only must the necessary competence be available to the person performing the task but also the associated responsibility must be transferred from the delegator to the person performing the task.
- In addition, mutual trust between the respective protagonists on a personal level is the essential basis for a successful path to self-organisation (Horsager 2013).

4.3   Viable, Lean, … and What About Agile?

While working on this book, the authors naturally also thought about agility – a topic that there is currently no getting around when dealing with companies and their organisations. With their deep conviction and enthusiasm for the viable system model (VSM), how should the authors position themselves on this? And above all, how can a model that was first formulated in 1959 provide an answer to current issues?

The authors are quite sure that the VSM – to name just one example – has always been agile! This becomes incredibly transparent when looking at the tasks of system 4: the constant outward view to customers, competitors, and partners; market events; technology trends; societal changes; changing expectations of companies; the demand to assume social and environmental responsibility; and various other concerns from other interested parties.

This conscious observation sharpens the company's understanding of the need to react adequately to these changes to ensure the company's long-term survival on the one hand and be able to position itself advantageously in competition on the other. An important instrument here is also the communication channel between system 4 and system 5, which can ensure that necessary organisational changes also find their way into the culture and self-image of the company, which is essential for a change process if it is to be successful.

Thus, the VSM has basic instances, communication channels, and processes for responding to change already laid out in its DNA. In addition, the VSM as an evolutionarily conceived management model results in an invaluable further advantage about the demand for agility: the general ability to adapt to changing environments and the constant review, readjustment, or selection of modifications made to achieve the best possible design for the company.

However, Fredmund Malik highlights this advantage of the VSM most clearly in the foreword to Sebastian Hetzler's book *Real-Time Control for Mastering Complexity*:

> Mastering complexity and thereby having the logic of evolution as an advantage on one's side makes the management of and in complex systems almost a pleasure.

(Hetzler 2010)

Further, the subject of agility intersects with that of lean in many areas, allowing us to determine many of the methods that it informs. For example, a pronounced customer orientation can be found in the context of agility, which is also explicitly required in lean, but also implicitly, for instance, in the customer cycle, pull instead of push, and one-piece flow, to name just a few. Agile and lean also share the human-centred self-image and the associated leadership culture if the latter is interpreted according to its original intention. This agreement may stem from the fact that both approaches require a particular culture, which as a result produces a certain mindset regarding the topics of employee management and development.

In particular, however, the image of people in the two approaches and the degree of congruence between them are also the subjects of very controversial discussions, since many people view lean more as an efficiency tool and therefore attribute less attentiveness to dealing with people to the approach. Here, it remains to be clarified whether the fundamental problem is not instead to be found in our Western (mis)interpretation of the topic of increasing efficiency through lean. In any case, the coaching kata in the lean context places the employee at the centre.

Of course, the authors do not want to hide these differences and controversial views. But in times of increasing division in all areas, which will undoubtedly prove to be, rather than goal-oriented, destructive about all current and future problems, the authors would instead emphasise the commonalities and a solution-oriented perspective.

In summary, the authors conclude that both a system-oriented approach (here in particular the VSM) and lean thinking offer many approaches, methods, and tools that can be found in the topic area of agility, or at least can be interpreted in terms of agility.

4.4 Digital Transformation

In the digital age, most information is available in digital form. Digital devices, which are faster, more powerful, cheaper, and more frequently connected every day are now ubiquitous. The global network of things and people make it possible for anyone to communicate with almost anyone else in the world. Cheap storage and intelligent access tools connected via the cloud enable easy storage and fast access to all information from anywhere.

This paradigm shift can be easily comprehended with the following drivers.

The drivers of success in the industrial age were:

- Ownership and control of the entire business process from the development department to the sales department.
- Top-down hierarchies and decision-making processes.
- Dominating use of company know-how.

The drivers of success in the digital age are:

- Use process components of others to respond quickly to changing customer needs.
- Self-organised, collaborative, cross-functional teams with a high degree of decision-making capabilities.
- Rapid learning from all interactions and applying new insights.
- Information is available in real-time to all demand groups regardless of where it originated.

It must also be mentioned that the benefits of a product can be enhanced for the customer by making associated and supplementary data available (e.g. via accompanying apps). On the other hand, the manufacturer gains exclusive information about its customers and their user behaviour through the data generated by these services, which can be used to optimise products and services.

The success factors in the functional hierarchy of the industrial age came from the exclusive work within the limits of their function, unity, and organisation. The three success factors of the digital age are manifested by the fact that they work within and outside the boundaries of their function, unity, and organisation.

The functional organisation, as it was created in Henry Ford's time and functioned in a contemporary way with its 'inside → out' value stream, was evolutionarily developed from the matrix and alliancing structures to a value network. The value network's organisational form uses the digital age's feedback density to optimise stakeholder value by placing the customer at the centre of decisions and actions. Thus, the insights gained from customer proximity can be addressed, and the work is based on an 'outside → in' value stream.

Governance and self-organisation are closely linked to digital transformation.

The digital transformation is subdivided into four categories:

- Digital data collection and analysis.
- Automation: autonomous/self-organising systems.
- Networking: connection and synchronisation of systems and activities that have not been networked before.
- Digital access: mobile access to the Internet and internal networks.

In industrial production, this involves strong product individualisation under the conditions of highly flexible (large-scale) production. The automation technology required for this is to become more intelligent through self-optimisation, self-configuration, self-diagnosis, and cognition and is able to better support people in their increasingly complex work.

4.5   Phases of Digital Change

Traditional business tends to be linear. Digital business is not linear. Today, companies need to look further and think broader. Their competitors are no longer just other companies in the industry. Their competitors are companies that see the opportunity to fundamentally change the industry to provide consumers with unprecedented products or services.

According to Venkat Venkatraman of Questrom Business School Boston, three competitors are competing against each other in every industry today (Venkatraman 2017):

- Industry leaders (traditional companies like Hochtief, VINCI, and Strabag).
- Tech entrepreneurs (relatively new companies using digital technologies to change the status quo, e.g. ICON or Ceapoint Desite).
- Digital giants (large companies whose core business is digital, e.g. Google, Autodesk, and Amazon).

Digital change can be divided, according to Venkatraman, into three different phases of transformation. These phases will be experienced by any company in any digital economy industry. They are:

- **Phase 1: Experimentation at the edge**
  This phase scans industries to see how and where other companies use digital technologies. Companies collect ideas, analyse them, and conduct initial tests. Nothing is changed in the core business.
- **Phase 2: Collision at the core**
  In this phase, the options from the experimental phase are limited to a few business processes that can generate real value. These are established in the company.
- **Phase 3: Reinventing the root**
  In this phase, the company undergoes a complete digital restructure. Old borders dissolve and new business models emerge at the interface of traditional industries. Systems that control change are digital.

Each competitor can be in different phases of change at any given time.

4.6 Digitalisation in the Construction Industry

The leading expert in artificial intelligence (AI) and chief engineer of Google Ray Kurzweil (2012) predicted that technical singularity will be reached by 2045, and this will completely restructure human life. 'Technical singularity' means that, because of the exponential progress of innovation and the resulting technologies, the future of technological development is no longer predictable. The thesis is highly controversial. Nevertheless, it is a fact that in the age of digitalisation the production industry is continuously confronted with new technologies.

Digitalisation will produce disruptive technologies in the construction industry and thus revolutionise elementary processes and techniques. In real terms, this means a reorientation of the design, execution, and in-use phases of construction projects. The following technology trends will have a major impact on the industry:

- Big data
- cyber physical systems
- modelling and simulations (e.g. building information modelling, or BIM)
- virtual reality (VR)
- augmented reality (AR)
- mixed reality (MR)
- blockchain
- artificial intelligence (AI)
- automatisation
- robotics
- 3D printing
- geoinformation systems and 3D laser scanning
- drones.

Many of the technologies are not yet ready for the market. A market study by Roland Berger (2016) shows that the technologies with the most significant added value are still in the development phase. This further development, driven by the aforementioned trends, can be described in different phases, starting with the current status and ending with fully automated and intelligent project management (see Figure 4.2). The phases can run in parallel, and specific subsectors in the construction industry are expected to develop faster than others.

The current status and the subsequent phases according to our classification are:

- Status quo.
- Phase 1: BIM, VR, AR, MR.
- Phase 2: Intelligent project management.
- Phase 3: AI in construction.
- Phase 4: Autonomous project management.

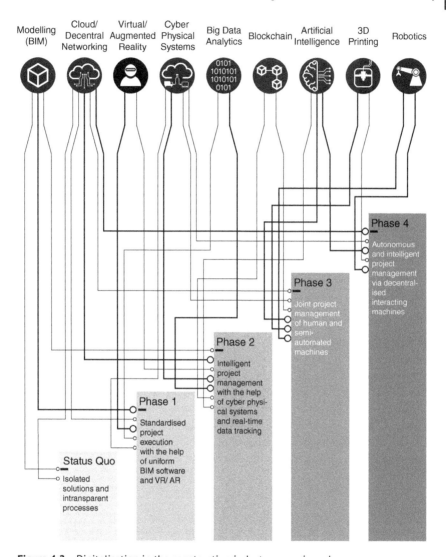

Figure 4.2 Digitalisation in the construction industry: overview phases.

4.6.1 Status Quo

According to Deutsche Telekom's (2019) digitalisation index, the small and medium-sized construction industry was ranked second last in 2017. Deutsche Telekom's realisation that the construction industry still has considerable deficits is confirmed by a survey conducted by Bearing Point management consultants.

It concludes that construction planning will not be able to satisfy future require-ments and that BIM[1] in Germany is in its initial phase compared to other countries.

The auditing firm PWC has found that only one out of 10 companies in the con-struction industry uses BIM (PWC et al. 2018). Two-dimensional plans continue to dominate the design process in the construction industry. There is currently no uniform standard about software usage, meaning various isolated solutions with limited compatibility are available. Currently, most 3D modelling uses insufficient detail of the digital model of construction with little or no linking of information (e.g. costs and deadlines).

It should be noted that BIM is a big step in the right direction. However, it is not the panacea for resolving all conflicts in the construction industry. The authors see the solution also in terms of partnerships and process orientation during project planning and execution. But one method or approach does not exclude the other. The approaches can be combined well, as the integrated project delivery (IPD) shows, for example (see Section 4.8).

4.6.2   Phase 1: BIM, VR, AR, MR

The first phase aims to unlock the potential of BIM fully (see Figure 4.3). This means that all project participants work in one model. It is located in a data-protected cloud. It can be accessed from anywhere and all relevant information (3D model, terms, costs, and information for facility management) is available. Development, design, approval, tendering and contract award, execution, handover, and operation using the digital model.

The model allows quick adaptation of parameters. The BIM model can be used as a configurator based on the automotive industry: the customer can choose from

---

1 BIM (building information modelling) is a digital method based on three- to n-dimensional object-oriented building models. The digital model of the building serves as an information source and data hub for the collaboration of the project participants. At the centre of this is the digital registration and interconnection of all relevant data for illustrating the physical, functional, cost, and time-related characteristics of a building. The usage of the data model and the continuous data update consider all project phases from survey until operation. The data can cover the entire lifecycle of the project and be used across all phases. If the BIM method is applied correctly and intelligently, added value can be achieved due to increased planning quality and improved schedule and cost reliability. This results from consistent information management and a collaborative and transparent working method. BIM is divided into different maturity levels (3D–6D): A 3D model means a 3D CAD (computer-aided design) drawing located on a common platform. Four-dimensional models contain additional information for scheduling, and 5D models add the cost of all components. Six-dimensional models are relevant for the usage phase and collect information for facility management.

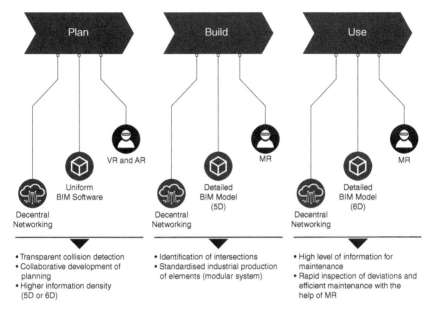

Figure 4.3 Digitalisation in the construction industry: phase 1.

several standards (keyword: modular system) and individualise their project. Standard elements can be produced in a production hall and delivered to the project.[2]

Another significant milestone will be the virtualisation and fusion of the virtual and real worlds. VR (e.g. Oculus Rift and HTC Vive) is a technology for immersion in a computer-generated world. Designers can design more realistically, and customers can formulate their requirements more concretely. Collisions between different designers become more visible and can be avoided.

AR[3] offers a direct or indirect view of a physical, natural environment. Elements are supplemented by virtual, computer-generated sensory inputs (sound, video,

---

2 The start-up company Katerra from Silicon Valley combined innovative construction methods with series production. The idea was that customers could configure their projects from standard elements. Production takes place in a production hall. More significant finished parts are then delivered to the construction site and assembled just in time. Even though Katerra went bankrupt in 2021, it was a good approach, and the industry learnt from them. The physical construction industry works differently from the private stationery industry or IT. Nevertheless, the construction tech scene continues to develop, and there will undoubtedly be many exciting and disruptive approaches.
3 Microsoft HoloLens is an MR eyewear that allows the user to display interactive 3D projections in the direct environment with the support of a natural user interface. The glasses work independently without a computer or smartphone. Trimble, the leading manufacturer of GPS technology, and the University of Cambridge are researching its use in the construction industry. Trimble has developed the SketchUp Viewer, which allows holographic 3D models to be projected into the real world.

graphic, or GPS data). Users can visualise site constraints and identify collisions, overlay times (from 4D BIM platforms), and cost information (from 5D BIM platforms) to generate optimal and efficient operations during design, execution, and operation. The technology can be used extensively for stakeholder information, occupational safety, operation, and maintenance in the future.

MR enables the precise alignment of holographic data on the construction site. Models or other information can be superimposed and anchored in the context of the physical environment. Trimble has developed a mass-produced construction site helmet with an MR visor for $300. This allows future construction supervision to quickly compare actual and target data for reporting and control purposes using the design data and the real construction progress.

It can also carry out regular visual inspections of building structures during operation. This is done using completely textured, geometrically exact models. The element surface texture is extracted from high-resolution images, allowing changes or defects to be detected automatically.

AR, VR, and MR applications are used in all phases of the product lifecycle and beyond in the mechanical engineering sector. For example, for:

- Layout planning in the company.
- Detecting potential problems with the product as early as the design phase.
- Advising the production department on the execution of machining processes.
- Simulation of set-up and clamping situations as well as the entire machining process.
- Training and instruction purposes.
- Support of the workers in maintenance tasks.
- Guidance for self-help by specialists not on site.

4.6.3   Phase 2: Intelligent Project Management

The next step of digital transformation in the construction industry is networking intelligent systems by cyber-physical systems (CPS) and developing AI in project management (see Figure 4.4).

In 2006, Helen Gill, Director of the United States National Science Foundation, defined the term 'CPS'[4]. CPS systems comprise physical, biological, or constructed systems that are integrated, networked, or controlled using information technology systems. Networking elements and evaluating information in real-time enable

---

4 CPS is also often used as a synonym for the Internet of Things (IoT). IoT is an extension of the Internet and can be described as any 'things' network. The physical and digital worlds are merging. IoT and CPS are very similar, but they come from two different areas. IoT was created by computer science and Internet technologies and focuses on the network idea. CPS has an engineering background and focuses more on communicating and cooperating with intelligent physical systems such as machines.

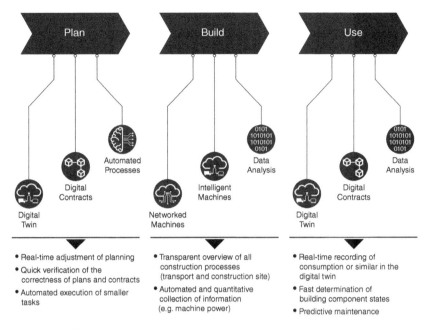

Plan	Build	Use
Automated Processes	Data Analysis	Data Analysis
Digital Contracts	Intelligent Machines	Digital Contracts
Digital Twin	Networked Machines	Digital Twin

• Real-time adjustment of planning	• Transparent overview of all construction processes (transport and construction site)	• Real-time recording of consumption or similar in the digital twin
• Quick verification of the correctness of plans and contracts		• Fast determination of building component states
• Automated execution of smaller tasks	• Automated and quantitative collection of information (e.g. machine power)	• Predictive maintenance

Figure 4.4 Digitalisation in the construction industry: phase 2.

the intelligent control of production processes. Machines or building parts become 'smarter' with the help of sensors (e.g. radiofrequency identification or localisation, motion, humidity, heat, pressure, or vibration sensors). Owing to powerful Internet connections, the miniaturisation of computer chips, cloud computing, and falling manufacturing costs for sensors, the IHS research institute estimates that 75 billion devices will be connected to the Internet by 2025.

CPS forms the basis for many other technologies that will add value to the construction industry. A first successful example of the initiation of intelligent construction sites was the SmartSite project of the Federal Ministry of Economics and Energy of Germany. All the machines used to build a road were networked via a cloud. The resulting information was collected using algorithms, evaluated, and used to optimise the construction process. Networking and communication between the machines and the cloud-enabled capacity meant that supply bottlenecks could be avoided and the traffic situation constantly monitored.

Hence, it is possible to develop a digital twin of the construction project. A digital twin is a virtual representation of a process, product, or service. The physical and virtual objects are connected in almost real-time with the help of sensors and data integration. SAP's senior vice president of IoT Thomas Kaiser believes that digital twins are essential to compete in the future market.

The digital twin covers all products, processes, and the plant's entire lifecycle. This applies to planning, execution, and, in particular, the usage phase. The condition of components and technology can be checked without further inspections and then mapped in the virtual model. Facility managers can collect, analyse, and evaluate data for maintenance from their workstations. Incidents can be predicted and avoided (keyword: predictive maintenance).

Another important topic is the use of blockchain to design intelligent contracts or smart contracts.[5] These are web-based contracts whose contract conditions are defined using algorithms. The contracts can be continuously checked and automatically reviewed as to whether contractual conditions have been fulfilled. It is possible to link a smart contract with others and thus generate multiparty arrangements for cooperative project management. These are automatically monitored via the blockchain.

Phase 2 is characterised by another technology trend, namely AI. AI can be used for various tasks in project management. Recurring tasks such as creating reports, printing plans, or ordering materials can be carried out independently by AI. People can concentrate on more creative and value-adding activities. The AI can use algorithms to define data-based recommendations to meet the project goals (keyword: recommendation engine).

AI still plays a small role in Phase 2. In the following phase, it plays a key role.

4.6.4 Phase 3: Artificial Intelligence in Construction

Phase 3 is characterised by cooperation between humans and AI. The machines and robots are extended by algorithms that enable them to take over human tasks (see Figure 4.5). In collaboration with the machine, humans will form a team that will allow a new dimension of productivity. Scientists Erik Brynjolfsson and Andrew McAfee (2014) from MIT even believe that the highest productivity can be achieved through the cooperation of human and machine.

One way to achieve this is to integrate AI and BIM. Designers can use the combination to create an optimised model. The design creator defines rules or criteria for the project initially, and under their consideration the AI can create proposals.

---

5 Blockchain is a digital database that lists and continuously expands the entire information history (e.g. for transactions). The blockchain was created by the development of the cryptocurrency Bitcoin and is the basis for the use of any cryptocurrency. Central institutions necessary for one currency (e.g. central banks) are replaced by several so-called miners. They check the correctness of the information in a decentralised manner and offer computing power for the digital database. The data is collected in blocks. Before a new block is created, the majority of miners must confirm the correctness of the block. The completed block is then concatenated with the previous blocks and these form the basis for the next block. Manipulations of historical data become almost impossible, since the blockchain would have to be changed on several thousand computers.

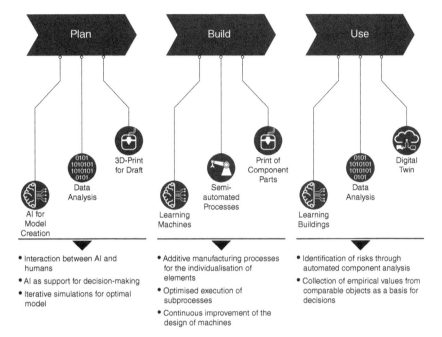

Figure 4.5 Digitalisation in the construction industry: phase 3.

These are based on data that the AI has previously retrieved from the network (e.g. component catalogues, reviews of new laws and regulations in building law, etc.). In addition, the AI can access an open database platform,[6] which will allow it to incorporate empirical values from similar projects that have already been completed. The AI supports the designer by indicating which buildings are nearby or how the immediate infrastructure can be optimally used. In addition, the AI can give designers direct feedback if one of their decisions leads to risks. Iterative simulations allow different scenarios to be played through in a short period.

AI in design is currently being researched and is already in use. Programs such as Dynamo[7] from Autodesk help designers with their decisions. The project Fractal has already optimised modelling with the help of intelligent algorithms. The algorithm determined an optimised spectator seat distribution during the construction of a football stadium.

---

6 In software development, it is standard to share knowledge. The idea behind this is to advance the entire development of a technology and to benefit from the experience of others.

7 Dynamo is an open-source platform for visual programming and an integral part of Revit. With Dynamo, you can create complex geometries, build parametric systems, perform calculations, bi-directionally synchronize data with external databases or access the Revit API and automate workflows.

Another example is the AI Alice, from the Californian company Alice Technologies, used for optimising schedules. In the first step, the scheduler feeds Alice with their expertise and defines several rules that the AI should consider. She uses this expertise to optimise herself. Alice then simulates different scenarios, evaluates them in terms of time and cost, and provides the project team with different variants for an optimal schedule. If unexpected factors are added during the project, Alice can communicate them, allowing schedulers to reschedule the program and identify opportunities and risks.

Conbrain, an Austrian company, offers various products for better project management based on AI. One of them is Early Bird, an AI-driven real-time risk management system that can help project managers manage risks as they arise. For this purpose, all documents used in the project – such as plans, emails, orders, offers, protocols, and contracts – are evaluated. Another interesting product from Conbrain is Smart Moodz, which analyses these documents according to emotions. This is an exciting application in the context of cooperative project development.

One of the essential characteristics of AI is that these systems are capable of learning. One of the methods of how machines can learn is machine learning. The machine learns by being trained with examples and then generalising them. The generalisation is made by pattern recognition so that the machine can also categorise unknown data (learning transfer).

Machine learning will play an essential role in the cooperation between humans and AI, as it helps them understand what is right or wrong on the one hand and learn humanoid action on the other. In certain areas, this machine learning is already so advanced that AI can already replace humans here, such as in the visual diagnosis of malignant skin lesions (Brinker et al. 2019).

One possibility for semiautomated construction sites is additive manufacturing methods.[8] Three-dimensional printing will play an important role here. It is possible to preprint all elements of the construction project in advance in a production hall using a 3D printer. Another scenario is the on-site printing of the building with the help of mobile 3D printers. The printer receives all of the relevant information from the BIM model, produces many different shapes, and thus individualises many products. Rooms can be created, which would not be possible using only humans. Components can be produced much more easily, quickly, and cheaply by printing the elements with honeycomb or grid structures and not filling them. Various materials (concrete, steel, or a cost-effective hybrid mixture) can be used as printing materials.

The start-up Apis Cor has already built an entire house from concrete using 3D printing within 24 hours.

---

8 With the additive manufacturing process, the entire object is successively created automatically with the aid of a robot. The robot receives its information directly from a 3D model and carries out the production independently using chemical and physical processes.

Another example is the US company Contour Crafting Corporation led by Professor Behrokh Khoshnevis's invention of the counter crafting process, which uses fully automatic portal robots to print buildings with a mixture of special and ordinary concrete based on digital data. This reduces the construction time of buildings to just a few days or even hours. Depending on the model, the construction printing robot for the first commercial generation has a working width of 8–12 m, weighs approximately 400 kg, and requires two operators. Another fascinating example is the Hadrian robot from Fastbrick Robotics. Hadrian can lay 1000 bricks per hour.

Even though the performance of the machines is remarkable, AI needs people in many ways. Owing to the high interdependencies and the high degree of participation of people, they are still required as principal actors. At present, machines can only solve specialised tasks more efficiently than humans can.

Nevertheless, AI benefits from cooperation because it learns from people and then advances to the next level: autonomous project management.

4.6.5   Phase 4: Autonomous Project Management

In the fourth phase, AI and machines can complete projects autonomously (see Figure 4.6). The human only expresses their wishes initially and defines variables that the AI needs to work out.

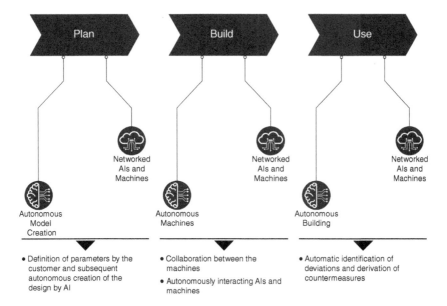

Figure 4.6   Digitalisation in the construction industry: phase 4.

Once the information has been received, various designs are modelled. The human being expresses themselves about the variants and adjusts their requirements. The AI takes the feedback and develops a model that meets the customer's expectations. It can then directly initiate all other processes. Thus, the AI automatically submits approval applications and orders all necessary materials.

The construction of the building now takes place autonomously via various systems. Autonomously moving means of transport communicate with other machines on the construction site so that the crane knows precisely when to unload materials, for example. At the construction site, different types of robots interact with each other. The specialised machines have defined the precise requirements to carry out their work. This catalogue of conditions is passed on to the robot that performs the work beforehand so that they will be aware of subsequent trader needs. Networking enables autonomous machines to warn each other of changes and adapt their activities to the new conditions accordingly.

In addition, intelligent building materials consisting of natural raw materials and robotics are used. With the help of nanotechnology, the material's properties can change and heal themselves in the event of damage.

The production method will change fundamentally due to learning autonomous machines and their associated AI. Construction production will constantly improve. There will be hardly any waste.

After the construction project has been completed, it receives an AI networked with the building parts and technology via the cloud. The object itself can determine its state of functionality, evaluate it, and determine when it needs repairing. If people agree, the building automatically requests the desired service.

In this futuristic picture, the human being plays a different role than before. Administrative and technical tasks are ultimately taken over entirely by machines. People will focus on the machines to educate them until they can develop themselves further.

Examples of autonomous communicating systems already exist. RWTH Aachen University has won the RoboCup Logistics League several times. In this competition, teams with several robots compete against each other. Their task is to manufacture simple products. The robots from Aachen were able to communicate autonomously using cooperative algorithms and were able to solve unexpected problems together.

4.7 Changing the Game

4.7.1 Nudge Management

Companies from Silicon Valley, such as Apple, Google, and Amazon, recognised the benefits early on of creating a culture of creativity and innovation for their employees. The behaviours of employees could be influenced by actively shaping their environment. The goal was to proactively consider employee instincts alongside the classic management tools that rationally affect employees' minds.

Richard Thaler, the winner of the Nobel Prize in Economics and a professor at the University of Chicago, recognised that small nudges stimulate instincts. The nudges form the framework conditions and ensure that the employees act according to the desired culture without thinking about it. What is important here is that the employees have freedom of choice and do not feel that they are obliged to do something.

Nudges can be divided into digital, physical, and social nudges; some examples are given here:

Digital Nudges

- Appropriate software and hardware that motivates you to work effectively and efficiently and supports you in making the right decision (e.g. gamification).
- Employer-initiated interactive fitness programs that remind employees of their balancing exercises online.

Physical Nudges

- Attractive work environments to promote knowledge sharing between colleagues and job satisfaction.
- Opportunities for chance encounters that lead to different perspectives and unexplored connections.
- Bar tables, whiteboards, and flipcharts in the meetings instead of overly comfortable armchairs.
- Calorie consumption signs in stairwells and by service bicycles.
- Lots of nature and plants that reduce stress and promote innovation.

Social Nudges

- Promoting interaction and culture as well as support groups and engagement.
- Reduce meetings, e.g. from one hour to 45 minutes.
- The company pays for childcare costs.
- Targeted promotion of female managers, free provision of fruit, personal thanks from the boss for good performance.

4.7.2   Tit for Tat

Suppose you internalise that the Toyota Production System has been used and developed for almost a hundred years. In that case, the state of optimised processes in the construction industry today is sobering. The framework conditions for joint overall process optimisation are not in place.

But it's not just companies in the construction industry that are experiencing difficulties with the sustainable implementation of lean principles. From a study of the Helsinki University of Technology (Mahlamäki et al. 2009), the following application challenges arose, among others:

- Distinguishing which activities add value during execution and which do not.
- Missing front-loading[9] for the prevention of fire-fighting actions.
- Missing personnel positions in an organisation (e.g. lean coordinator).

Alan Mossman (2016), the co-founder of the lean construction movement and the last planner system, sees the limited performance of construction projects, particularly in project complexity and unique production. If one compares the development of labour productivity in the German construction industry with other sectors, productivity from 2002 to 2013 grew on average by about 0.2%. The pharmaceutical industry, for example, is growing at an average rate of 4.2%.

An essential prerequisite for the successful application of lean is a common culture. The basis for this is trust, which must be earned. To gain trust, a principle from game theory is recommended:

tit for tat, which means: 'Like you to me, like I to you.'

The renowned political scientist Robert Axelrod proved this with his much-acclaimed experiment.[10]

Tit for tat can be attributed to four factors:

- Cooperation requires friendliness.
- Attempts at exploitation must be retaliated against.
- Indulgence should take place when the other signals readiness to cooperate again.
- Own behaviour must be easily predictable.

With tit for tat, you always start with cooperation and adjust to the other person. An adaptation of the basic strategy is, for example, 'tit for two tat,' whereby one, despite exploitation twice, remains cooperative. Axelrod's strategy is a superior cooperation strategy that always wins in the long run.

---

9 'Front-loading' means to plan and to take measures for events that are foreseeable, so that arising problems can be solved immediately without getting into a stress situation (e.g. suitable public relations work before the start of construction measures towards those affected).
10 In his book *The Evolution of Cooperation*, Axelrod (1984) illustrates the strategy tit for tat using a two-person competition at a computer tournament. The experiment shows that cooperation can arise independently of an external instance (e.g. laws that lead to collaboration) even among egotistic individuals.

Trust plays an essential role in realising construction projects. According to a survey conducted by Autodesk in 2020, organisations with the highest level of trust report millions of dollars' worth of benefits, such as:

- Lower voluntary turnover that would otherwise be spent on replacing staff.
- Fewer missed schedules, resulting in gains of up to $4 million a year.
- Higher levels of repeat business, driving gross margins up to 7% higher.
- People are twice as likely to be explicit about their requests, compared to organisations with average trust.
- Managers are twice as likely to share feedback and are more likely to develop their employees.

Hence, 43% of the highest trust firms make collaboration central to working with other organisations across projects. This results in benefits such as:

- Share information openly and easily (single source of truth).
- Receive responses to queries more quickly.
- Hear about problems faster.

4.8 Partnering

In the 1990s, the British petroleum company BP plc faced the great challenge of finding an economical way to exploit new oil reserves. The only possibility was to reduce its high development costs. The solution was for the various oil companies to form a contractual alliance and share their knowledge. As a result, the original target costs were 36% lower, and the project was completed three months ahead of schedule. The conclusion is that the cooperative and partnership approach achieved a common goal that the parties would not have achieved alone.

Partnering models or alliances in the construction industry mean early involvement of the parties needed (e.g. general contractors, technical consultants, designers, specialist trades) by the client. The idea is that all responsibilities, opportunities, and risks are shared collectively during the realisation of the project: 'Pain and gain', i.e. all positive and negative events are shared with all project participants. Everyone bears the danger they can take or which is assigned to them.

Project participants thus become project partners. Bunker mentalities and isolated ways of thinking are replaced by a culture that makes trust, collaboration, innovative solutions, and honest and open communication (no hidden agendas) possible. An open-book policy applies, information is shared, and the target costs are checked and only adjusted in consultation with the client. The implementation interests of the parties involved are aligned.

Partnering models are suitable for all projects, but especially for large or megaprojects, which are difficult to define at the beginning and have a long duration with many unknowns.

Countries such as Australia, Great Britain, the United States, and Finland have already achieved significant successes through partnering.

With the ACA Project Partnering Contract (PPC 2000)[11] developed by the Association of Consultant Architects Ltd. in cooperation with David Mosey of King's College London, the UK has achieved cost savings of up to 30% on selected projects over 10 years. A multiparty contract was concluded with the project participants (client, specialists, consulting engineers, planners, general contractors, etc.) for two project phases. This was divided into the preconstruction phase (preparation, design, risk analysis, etc.) and the construction phase (execution).

According to Ailke Heidemann (2011) of the Boston Consulting Group, the advantages of a multiparty contract are the avoidance of interfaces and a more vital link between those involved in the project. There is clarity between the contracting parties regarding who has what role and what area of responsibility through the whole production system.

The following shall be specified in a partnering agreement:

- An early warning system to identify possible problems during design and execution.
- The establishment of a core group that makes decisions.
- A common schedule and cost and risk management.
- Incentive mechanisms (bonus/malus).
- A clear decision regulation for the dispute and a partnering consultant who is available to advise the core group.

The integrated project delivery (IPD) model from the United States is another variant of a multiparty contract. Instead of the usual clash of single interest contracts, a contractual model is developed, forming the project organisation into a community of interests. The client, planner, and contractors define the qualitative, cost, and time objectives. If these are met, everyone benefits from a bonus. On the other hand, if they are not achieved, a penalty is due for all of them. In contrast to other models, IPD combines partnership contract models with lean design or construction and/or BIM.

---

11 It is also worth mentioning the PPC International (Project Partnering Overseas), as well as the TPC 2005 (Team Partnering Contract) and the SPC 200 (Specialist Sub Contract) for the integration of subcontractors. Another British multiparty contract worth mentioning is FAC-1 (Framework Alliance Contract-1). This is a framework contract for alliance formation. Also worth mentioning is the Australian Project Alliance Agreement with its focus on the no-blame culture and the American Document C191-2009 from the American Institute of Architects. This list does not claim to be exhaustive.

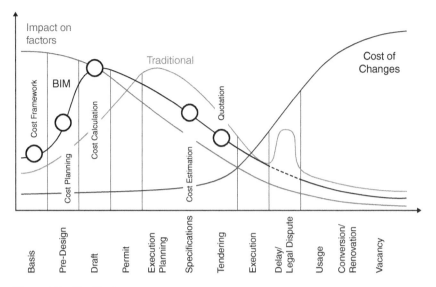

Figure 4.7 The MacLeamy curve.

What these approaches have in common is that they require the parties involved to coordinate and define the decision-making process and the issues to be regulated in planning and implementation at a very early stage. This means that the chances of influencing quality, cost reduction, meeting deadlines, or achieving an exit from the project are much improved.

See Figure 4.7 by Patrick MacLeamy, founder of BuildingSMART International.[12]

By international standards, the German construction industry is still in its infancy.

Large private construction projects can be carried out in two-step models based on guaranteed maximum price (GMP) contracts[13] with 'cost plus fee'[14] elements. The implementation of private projects is flexible, particularly in terms of financing, and is already being used by the major players in the construction industry.

---

12 MacLeamy established a concept, commonly referenced in the design and construction industry as the MacLeamy Curve, to illustrate the escalating cost of design modifications as a project team progresses in the design process.
13 According to Shervin Haghsheno (2004) of KIT, a GMP contract is a construction contract, in which the payment for the contractually owed construction work is based on the actual manufacturing costs incurred plus a surcharge for general business costs as risk and profit. The remuneration may not exceed a contractually agreed target price (GMP).
14 This is a cost/reimbursement agreement. The contractor discloses their cost estimate. On this basis, the client pays a percentage (fee) on the costs associated with the construction of the building. In earlier versions of the VOB/B, there were cost-plus-fee contract models, but these did not become generally accepted in the application.

On the other hand, projects financed by the public sector continue to be tendered throughout Germany by the VOB/A (German Construction Contract Procedures) and throughout Europe, i.e. by the sector's directives. This, unfortunately, means in many cases that the cheapest provider and not the 'best for project' applicant/team wins the contract.[15] The public sector is aware of the problem, and the federal states initiate contract models that advocate a partnership-based approach, as shown by various pilot projects (ECE, Hamburg Port Authority, Partnerschaftsmodell Schiene of Deutsche Bahn) of private and public clients. Both science and practice have recognised the potential of the approach and will establish these methods.[16]

The possibilities of partnering models in combination with a collaborative culture will immensely strengthen the application of lean construction and BIM, as synergy effects are created and a legal framework for implementation according to lean and BIM principles is available.

4.9 Success Patterns in Projects

The considerations in this chapter have essentially arisen in connection with megaprojects but are also valuable and worthy of attention in implementing projects of any size. The only thing to watch out for is the correct dose in terms of effort.

In the past, most reporting and academic literature has dealt with the poor performance of large-scale and megaprojects. That means you know at least part of the reason why something doesn't work. But what about the projects that were a success? If one fully believes Flyvbjerg's Iron Law of Megaprojects (2014), at least every 10th project should be a success; if one follows Merrow (2012), one-third of the projects are a success in the context of the Iron Triangle. For this reason, it seems helpful to look now less at the causes of failure and more at the patterns of success. This chapter, therefore, looks at supposedly successful projects in the context of the Iron Triangle and at relevant literature and presents the findings.

When dealing with large-scale and megaprojects, the same questions are always asked: How do successful projects work? What is the 'secret' recipe for success? What is the simple solution for solving the complex task?

The authors did not find simple solutions, but there are some points in which the 'successful' projects are similar. Here one can recognise regularities, repetitions, similarities, or even patterns. The following projects were considered

---

15 Since review procedures are time-consuming and costly, the focus can be on highly formalised and price-oriented awards.

16 For example, the IPA Zentrum (www.ipa-zentrum.de) German centre for IPD and the GLCI (www.glci.de) German Lean Construction Institute.

(Project Database: Metro Madrid, Oresund Bridge, Mall of America, Steinbühl tunnel, Heathrow T5, Guggenheim Museum Bilbao). On this basis and under-appreciation of relevant literature such as Merrow (2012), Greiman (2013), but especially Denicol et al. (2020), the formation of areas under which these points can be subsumed was made.

The areas are shown and roughly explained in the following:

- **Decision-making:** Here you can find topics that improve decision-making and ensure the 'debiasing' of decisions. Naturally, the freedom of decision and design is greatest at the beginning of the project, so the topic should undoubtedly find its importance here. Still, it is essential over the entire course of the project.
- **Organisational design:** Following Merrow (2012) or Greiman (2013), the organisational framework plays a vital role in success. Therefore, topics of organisational design are found in this area, which include structure, process, and control organisation in the context of the life cycle. However, the topics of management and personnel are also located here. Organisational design is an area that is challenging throughout the entire course of the project. The need for resources is probably greatest during the construction phase, which must be taken into account strategically.
- **Strategic orientation:** In this area, the strategy or the highest normative and strategic orientation is found, where topics such as security, transparency, coop-eration, and alignment of interests are placed. Setting the right course for a project from the beginning and staying on track is critical to success.
- **Risk and uncertainty:** This area includes, for example, the topics of fair risk distribution, risk minimisation through the use of proven technology, and the phenomenon of the black swan[17]. In the context of the course of the project, it must be taken into account that risks can be handled even better in early project phases.
- **Stakeholder management:** The essential stakeholder management is recorded in this area. Experience shows that this plays a decisive role in conveying abstract, complicated issues, especially before implementation. If one is with a project in the building, one can often inspire with the conversion in itself.

_____

17 In individual cases, for example, cost overruns of complex, major, or megaprojects reach extreme dimensions. Financial mathematician Nassim Taleb (2010) refers to such phenomena as black swans. These are unpredictable, rarely occurring outliers that have a disproportionate negative impact on the project. Black swans are the unknown unknowns that, despite all plans and precautions, still surprise an organisation and can lead to catastrophic consequences. Taleb sees it as useless to try to predict black swans. They are characterised by their unexpected appearance. According to Flyvbjerg and Budzier (2011), a black swan exists when the budget is exceeded by 200% and 70% more time is needed.

- **Planning:** Planning and preparation are located here. This often takes a very long time and accounts for two-thirds of the total duration in some projects. The key to success here is the very high planning depth in early planning phases and the effective management of changes and interfaces. In principle, the credo 'plan first, then build' should apply. Planning has its focus before implementation but plays a success-critical role over the entire course of the project.
- **Production:** If all of the above points have been implemented well, production is the most stable phase of the project. If not, it can be very chaotic. Experience has shown that it is favourable for production if the customer and the contractor work from the production site. There is a common production system that focuses on creating the product.

Based on the above database, 40 points could be identified and assigned to the previously explained areas. Many of these points are also relevant for 'normal' or smaller projects. However, the data basis used for the evaluation is specific to large and megaprojects. The result is not yet representative because the dataset is too small, but it nevertheless shows an initial tendency and invites further study of the topic. The areas and points are not exhaustive. There is no single promising approach. Instead, it is essential to have a pleasing interplay of many approaches and measures and their effective adaptation. This is also the conclusion reached by Denicol et al. (2020).

Decision-making

1. Pay attention to 'optimism bias'.
2. Pay attention to Hirschman's 'hiding hand'.
3. Pay attention to 'escalating commitment'.
4. Pay attention to 'strategic misrepresentation'.
5. Perform 'reference class forecasting'.
6. Carry out control by independent experts.

Organisational Design

7. Employ authorised and experienced managers.
8. Employ experienced project personnel.
9. Build a clear structure, process, and control organisation.
10. Ensure autonomy and cohesion (e.g. between programme and project).
11. Ensure rapid and appropriate staff development.
12. Do further development of the organisation along the lifecycle.

Strategic Orientation

13. Make safety a top priority.
14. Build a transparent and constructive error culture.

15. Create a 'one team' mentality to promote proactive action.
16. Express a clear willingness to cooperate and collaborate.
17. Ensure the alignment of interests.

Risk and Uncertainty

18. Use of modern but also purged technology.
19. Reduce the preparation phase (if planning depth is high).
20. Pay attention to black swans.
21. Establish a fair distribution of risk.
22. Ensure intensive consideration of costs, deadlines, risks, interfaces, and changes across all project phases.

Stakeholder Management

23. Note: public relations and press relations are an essential part of good project management.
24. Project information must be provided based on substantive planning.
25. Involve and inform key stakeholders at an early stage.
26. Major and megaprojects are a 'public journey', i.e. professional stakeholder management from the beginning is essential.
27. The limits of participation must also be identified.

Planning

28. Ensure early integration of implementation know-how.
29. Ensure high planning depth in early project phases.
30. Work with freeze zones in the planning process.
31. Build an effective management of changes and interfaces.
32. Work with cross-functional teams.
33. Apply a context-dependent project management method mix.

Production

34. Do not perform pure price allocations.
35. Apply standardisation, modularisation, and prefabrication.
36. Establish an effective and efficient production system (based on the preceding considerations, the authors recommend lean management, or a variant for this purpose).
37. Encourage collaboration monetarily.
38. Ensure rapid decision-making.
39. All parties should work from the point of production (i.e. at the construction site).
40. Disputes should be resolved out of court.

Since the question also arises about how these groups and points are related, the issues were connected and examined in social network analysis with Kumu.io.[18] The evaluation of the degree centrality of the Social Network Analyse yielded a subjective result with the following ranking:

1. Employ experienced project personnel.
2. Encourage collaboration monetarily.
3. Ensure alignment of interests.
4. Employ authorised and experienced managers.
5. Ensure rapid decision-making.
6. Express a clear willingness to cooperate and collaborate.
7. Establish an effective and efficient production system (based on the preceding considerations, the authors recommend lean management, or a variant for this purpose).
8. Ensure high planning depth in early project phases.
9. Use of modern but also purged technology.
10. Build a clear structure, process, and control organisation.

With the above 40 points, you have a framework or evidenced patterns of success to intelligently adjust projects regardless of size with valuable resources over the entire project lifecycle from idea to deconstruction.

In general, it would be desirable to exchange more information about successful programmes, projects, and subprojects, especially those assigned to large-scale and megaprojects, and examine and record the patterns of their success to test their adaptability. These success patterns are then located in the project's lifecycle. The understanding around and the handling of complexity and the transfer into an effective organisation model play an important role here.

4.10 Findings, Criticism, and Reflective Questions

4.10.1 Findings

As a manager, you should always strive for an organisational form that controls itself. You should create the appropriate framework conditions for this.

Digital change is divided into three different phases of transformation.

- Phase 1: Experimentation at the edge.
- Phase 2: Collision at the core.
- Phase 3: Reinventing the root.

---

18 https://www.kumu.io/MichaelF/success-patterns-in-complex-projects (accessed 22 March 2022).

Digitalisation in the construction industry can be divided into the following phases:

- Status Quo

  Isolated solutions and intransparent processes.

- Phase 1: BIM, VR, AR, and MR

  Standardised project management with the help of integrated BIM software and VR/AR.

- Phase 2: Intelligent project management

  Intelligent project management with the help of CPS and real-time data acquisition.

- Phase 3: AI in construction

  Joint project execution of human and semiautomated machines.

- Phase 4: Autonomous project management

  Autonomous and intelligent project management via decentralized interacting machines.

According to Richard Thaler, Nobel Laureate in Economics and Professor at the University of Chicago, instincts stimulate small nudges.

Tit for tat underlines the importance of cooperation and collaboration. It can be traced back to the following factors:

- Cooperation requires friendliness.
- Exploitation attempts must be rewarded.
- Indulgence should take place when willingness to cooperate is signalled.
- Own behaviour must be easily calculable.

Based on this, trust can be built up with the other project participants.

Partnering models enable the contractually joint execution of construction projects. The goal of client and contractor can be aligned, and an overall optimum can be generated for project implementation. Contractual opponents can become contractual partners.

Analysing and adapting patterns of success can be a smart approach.

4.10.2 Criticism

Unfortunately, digital transformation is not a sacred cow. The mapping, monitoring, and evaluation of processes must be taken into account so that before the digitisation of processes they are fundamentally reviewed and, if necessary, revised. If inadequate processes are digitised, the result will inevitably be inadequate digitised processes.

- As noted elsewhere in this chapter, digitisation and many of its specialist features require human support to function. In many cases, this requires specially trained personnel, who are either rarely available on the labour market or are correspondingly expensive.
- In addition to the dependence on specialised personnel, digital solutions also depend on the availability of a stable power supply and other trouble-free supply infrastructure. This can prove to be a limiting factor in the event of problems or, in the case of a blackout, can paralyse entire branches of industry.
- Because of this dependency and the in-house know-how used, these new systems are also particularly vulnerable to attacks from the field of cyber-crime, which is growing at exponential rate just like the technology it targets. Therefore, extensive security measures must be taken to protect internal data, which means additional expense. If these security measures fail, negative incidents can have a massive impact on the trust of existing and potential customers.
- In addition to these technical threats, there are also numerous reservations from a human perspective. For example, some of these systems pose more of a threat to humans due to their technological idiosyncrasies. This is, for example, the reason the working area of older generations of industrial robots has to be separated from the functional areas of human employees in a complex way and by expensive technological measures (e.g. by light grids or similar).
- Newer generations of industrial robots can now also be used in collaborative applications (hence the name 'cobots'). In some cases, there are reservations on the part of humans, which on the one hand can be attributed to a general rejection of the use of new technologies, as can happen in any change process. On the other hand, the 'uncanny valley' phenomenon has been described for the use of humanoid robots, which consists of the fact that human empathy can turn into disgust if the robot exceeds a certain degree of human appearance and behaviour. The phenomenon was first described in 1970 by Japanese robotics engineer and robotics professor Masahiro Mori (Kunze 2019). Since then, it has been studied worldwide.
- Finally, special care must be taken when generating, collecting, processing, and storing internal, personal, and customer data. In addition to protecting know-how, trade secrets, and customer secrets, personal data, mainly, are considered worthy of special protection in many countries. For this reason, there are a large number of national and international standards, norms, and laws (e.g. DSGVO in Germany, ISO/IEC 27001, TISAX), which in some cases severely limit the possibilities of what is technically feasible.

Cooperation and alignment of interests is a mighty lever for much better performance and effective complexity management. However, it must be wanted by all parties involved. Top management, in particular, must be serious about this and not pursue hidden agendas.

4.10.3 Reflective Questions

- Have you built your team to be redundant as a leader?
- Is there a motivating work environment for your team that encourages identification and independent work?
- How can identification, independent work, and monitoring be combined in a way that makes sense and is accepted by employees?
- In which phase of the digital transformation is your company, where does the company want to go, and what still needs to be done?
- How can reservations about digital transformation among employees be reduced (e.g. fear of job loss)?
- How can digital transformation and data protection (especially in the processing of personal data) be reconciled in the best possible way?
- Do you turn on 'tit for tat'?
- Is your environment or organisation ready for partnering? What can you do to establish partnering? For example, can you pilot a project?
- Which methods or approaches can be used to reconcile the steering of many with a predefined goal achievement in the best possible way?
- Do you have the success patterns in mind? Do you build your project along these patterns?

5

Summary and Closing Remarks

All Models are wrong, but some are useful

George E.P. Box

You should now know how to approach the design of intelligent projects, and even though this book has focused on construction projects, much of what you have read can be generalised to other fields.

As described, 'intelligent' can stand for:

- Adaptability and robustness.
- Logic and sensibility.
- Low-waste processes.
- Use of new technologies.
- Consistent use of employees' skills and abilities.
- Cooperative collaboration.
- Sustainability and responsible use of resources.
- Production systems:
  - in which it is happy to work
  - in which there is a working culture of motivation and empowerment.

The following results are given if we subsume these requirements under the individual chapters.

5.1 Complexity, Cybernetics, and Dynamics

Complexity in a changing world is not a new phenomenon. The perception of it existed for ages before there was a term for it. However, understanding and dealing effectively with complexity has never been more important. The world has become fast; there are unprecedented amounts of data and information to be assessed, and the level of interconnectedness is advancing at an incredible pace. Many complex issues such as climate change, sustainability, social justice, and digital transformation have to be thought of together. This inevitably leads to extremely complex systems which one can only navigate by employing holistic thinking. Complexity can be understood as a dimension or fuel of systems. Chapter 1 introduces the understanding, handling, and some methodological approaches, especially the viable system model (VSM), which as a system approach is discussed in more detail. The authors believe this is a powerful model for understanding, analysis, and design. But also, the basic understanding of system dynamics with its archetypes helps to sharpen our view of the networked world.

5.2 Lean Management and Lean Construction

If you consider that lean management came into being at a time of scarcity, it takes on a completely different significance. There was not much of anything, and out of this necessity one had to make the best of it. That means lean management is all about creative thinking and doing. Also, this means, on the one hand, to regard continuous improvement as a never-ending process of constantly questioning the familiar and, on the other hand, to align oneself transparently with the customer's added value and avoid and eliminate waste. Around these basic ideas, a strong attitude approach and a large catalogue of tools and methods have been developed for all industries and specifically for construction projects. This leads demonstrably to more efficiency and productivity. In the construction industry in particular, however, the alignment of the interests of the entire system must be kept in mind. If only a single institution aligns its production according to lean and the whole system does not, this has little effect. With the practical reports of Martin Jäntschke about the introduction of lean in a group of companies and large projects, this second chapter moves from theory to practice.

5.3   Cybernetic and Lean

Chapter 3 shows that cybernetics as a general metascience and systemic approach can be combined very well with lean strategies and can help us to gain a deep understanding to implement systems, organisations, and projects effectively and efficiently. Also, the important performance measurement was shown based on different examples. The authors find the 'Beers' Triple' fascinating, which builds a bridge to the VSM and thus poses essential questions in terms of operational and strategic implementation and development. The case studies then offer a bridge to practise the various lifecycle phases of actual projects. The example of a medium-sized engineering manufacturer whose entire organisational DNA is based on the VSM will then need to be developed further.

5.4   Beyond Cybernetic and Lean

The topics of self-organisation and digitalisation are closely linked and, when considered together, offer great potential for developing greater efficiency and effectiveness. But also a fundamental understanding of these trend topics in the context of organisational and project system structures and how they interact is essential to be able to classify and effectively establish them. The phases of digitalisation we have outlined show a possible path. However, it is crucial to remain critical. Digitalisation is not the sacred cow for solving all problems; the motto is still: 'If you digitalise a bad process, you get a bad process.' Therefore, digitalisation also offers the opportunity to think about things in a new and different way.

Fundamentally, however, it is also about a paradigm shift in project execution, away from confrontational, wasteful implementation to orientation and alignment of the overall system towards the most intelligent and best project implementation. The topic of partnership, which is briefly touched upon in Chapter 4, offers excellent potential. Projects can become complex very quickly, and the literature and our own experience have shown that there is no single right way to successfully implement them. Instead, it is an interaction of many individual factors, which must be considered over the lifecycle of the project. System patterns are a good approach to understanding complex systems and forming practicable heuristics.

With this book, we have presented and linked old and new topics. Although system approaches or cybernetics sometimes seen unwieldy or impractical, we believe they can help to answer many questions that would otherwise go unanswered – and, indeed, unasked.

However, we are not the representatives of the one true approach and the one true solution. As the book shows, there is no such thing. Problem and solution remain context-dependent. Mental models help us classify and understand things, but common sense and our thinking are and remain potent tools to find our way in the world and face complexity. Or as the statistician George E.P. Box put it in his now-famous quote:

All Models Are Wrong, But Some Are Useful

This should help us to remember that the models we create are approximations to reality. They are not absolute truths but valuable tools to organise our living together. Yuval Noah Harari, professor of macrohistory at the University of Jerusalem, states that people can organise their coexistence although they lack any biological instinct. According to Harari, they achieve this through order and writing.

This book is a modest contribution to this.

References

Appelo, J. (2010). *Management 3.0: Leading Agile Developers, Developing Agile Leaders*. Addison-Wesley Signature Series (Cohn).

Ashby, R.W. (1985). *Einführung in Die Kybernetik*, 2e. Frankfurt am Main: Suhrkamp.

Autodesk (2020). Trust matters: the high cost of low trust. http://constructioncloud.autodesk.com/rs/572-JSV-775/images/High_Cost_of_Poor_Trust.pdf (accessed 22 March 2022).

Axelrod, R. (1984). *The Evolution of Cooperation*. Basic Books.

Bach, V., Berger, M., Henßler, M. et al. (2016). *Messung der Ressourceneffizienz mit der ESSENZ-Methode*. Berlin: Springer.

Ballard, G.H. (2000). The last planner system of production control. PhD thesis (dissertation ed.). Birmingham: Department of Civil Engineering, University of Birmingham.

Ballard, G. and Koskela, L. (2011). A response to critics of lean construction. *Lean Construction Journal IGLC – Special Issue* 2011: 13–22.

Beer, S. (1963). *Kybernetik und Management*. (trans. I. Grubrich). Hamburg: S. Fischer Verlag.

Beer, S. (1974). *Designing Freedom*. Chichester: Wiley.

Beer, S. (1979). *The Heart of Enterprise*. Chichester: Wiley.

Beer, S. (1985). *Diagnosing the System for Organizations*. Chichester: Wiley.

Beer, S. (1995). *Brain of the Firm*. Chichester: Wiley.

Bergbauer, A.K. (2008). *Six Sigma in der Praxis: Das Programm für nachhaltige Prozessverbesserungen und Ertragssteigerungen*, 3e. Renningen, Germany: Expert Verlag.

Berger, R. (2016). Digitalisierung der Baubranche. https://www.rolandberger.com/de/Media/Digitalisierung-der-Baubranche.html (accessed 22 March 2022).

Bertagnolli, F. (2018). *Lean Management*. Wiesbaden, Germany: Springer Fachmedien.

Designing Intelligent Construction Projects, First Edition. Michael Frahm and Carola Roll.
© 2022 John Wiley & Sons Ltd. Published 2022 by John Wiley & Sons Ltd.

Boateng, P., Chen, Z., Ogunlana, S., and Ikediashi, D. (2012). A system dynamics approach to risk description in megaprojects development. *Organization, Technology and Management in Construction* 4 (3): 593–603.

Brecher, C., Corves, B., Schmitt, R. et al. (2015). Kybernetische Ansätze in der Produktionstechnik. In: *Exploring Cybernetics* (ed. S. Jeschke, R. Schmitt and A. Dröge). Wiesbaden, Germany: Springer Fachmedien.

Brinker, T., Hekler, A., Enk, A. et al. (2019). Deep learning outperformed 136 of 157 dermatologists in a head-to-head dermoscopic melanoma image classification task. *European Journal of Cancer* 113: 47–54.

Brookes, N. (2015). *Delivering European Megaprojects: A Guide for Policy Makers an Practicioners*. Leeds: University of Leeds.

Brunner, F. (2011). *Japanische Erfolgskonzepte*. Munich: Carl Hanser Verlag.

Brynjolfsson, E. and McAfee, A. (2014). *The Second Machine Age: Work, Progress, and Prosperity in a Time of Brilliant Technologies*. Norton & Company.

Chakravorty, S. (2009). Process improvement: using Toyota's A3 reports. *Quality Management Journal* 16 (4): 7–26.

Complexity Management Academy – Abteilung Innovationsmanagement am WZL der RWTH Aachen. (2020). Informationsdokument für interessierte Unternehmen. Komplexitätsarbeitskreis. Aachen: WZL der RWTH Aachen.

Conant, R. and Ashby, W. (1970). Every good regulator of that system must be a model of that system. *International Journal of System Sciences* 1 (2): 28–97.

Conway, M. (2021). Conway's law. http://www.melconway.com/Home/Conways_Law.html (accessed 22 March 2022).

Cruikshank, R. (2019). So… what is a big room? https://leanconstructionblog.com/So-What-is-a-Big-Room%3F.html (accessed 22 March 2022).

Darwin, C. (1859). *On the Origin of Species*. London: John Murray.

Deming, W. (1986). *Out of the Crisis*. New York: Cambridge University Press.

Demir, S.-T. and Theis, P. (2018). Lean construction management (LCM®). In: *Lean Construction – Das Managementhandbuch: Agile Methoden und Lean Management im Bauwesen* (ed. M. Fiedler), 137–162. Munich: Gabler Verlag.

Denicol, J., Davies, A., and Krystallis, I. (2020). What are the causes and cures of poor megaproject performance? A systematic literature review and research agenda. *Project Management Journal* 51 (8).

Depiereux, P. (2021). 5 Fragen zum MVP. https://www.computerwoche.de/a/5-fragen-zum-mvp,3544544 (accessed 22 March 2022).

Design Methods Finder (2022). Steep analysis. https://www.designmethodsfinder.com/methods/steep-analysis (accessed 22 March 2022).

Deutsches Institut für Normung e.V (2015). *Qualitätsmanagementsysteme – Grundlagen und Begriffe (DIN EN ISO 9000:2015)*. Berlin: Beuth Verlag.

Deutsche Telekom (2019) Digitalisierungsindex Mittelstand 2019/2020: Der digitale Status Quo des deutschen Mittelstandes. https://www.digitalisierungsindex.de/wp-content/uploads/2019/11/techconsult_Telekom_Digitalisierungsindex_2019_GESAMTBERICHT.pdf (accessed 22 March 2022).

Dirnberger, H. (2008). *Eine Systemorientierte Managementmethode für Consulting Unternehmen im Bauwesen*. TU Cottbus.

Dombrowski, U. and Mielke, T. (2013). Lean leadership – fundamental principles and their application. *Procedia CIRP* (7): 569–574. https://doi.org/10.1016/j.procir.2013.06.034.

Dloughy, J., Binninger, M., Oprach, S., and Haghsheno, S. (2018). Mastering complexity in takt planning and takt control: using the three level model to increase efficiency and performance in construction projects. https://www.academia.edu/37629374/Mastering_Complexity_in_Takt_Planning_and_Takt_Control_Using_the_Three_Level_Model_to_Increase_Efficiency_and_Performance_in_Construction_Projects (accessed 22 March 2022).

Dominici, U. and Mielke, T. (2010). A viable system view of the Japanese lean production system. Proceedings of the World Complexity Science Academy First Conference: Complexity Systemic Science and Key Global Challenges of Our Times, pp. 1–34. https://www.academia.edu/2601712/A_Viable_System_View_of_the_Japanese_Lean_Production_System (accessed 22 March 2022).

Dörner, D. (2012). *Die Logik des Misslingens*. Rowohlt.

Drucker, P. (1977). *People and Performance*. Harper's College Press.

Drucker, P. (1969). *The Age of Discontinuity*. New York: Harper & Row.

Eco, U. (1988). *Il pendolo di Foucault*. Mailand, Italy: Bompiani.

Espejo, R. (1989). A cybernetic method to study organizations. In: *The Viable System Model: Interpretations and Applications of Stafford Beer's VSM* (ed. R. Espejo and R. Harnden), 361–382. Chichester: Wiley.

Fiedler, M. (ed.) (2018). *Lean Construction – Das Managementhandbuch*. Munich: Springer Gabler.

Flyvbjerg, B. (2014). What you should know about megaprojects and why: an overview. *Project Management Journal* 45: 6–19. https://doi.org/10.1002/pmj.21409.

Flyvbjerg, B. and Budzier, A. (2011). Why your IT project may be riskier than you think. *Harvard Business Review* 89 (9): 601–603.

Forrester, J. (1971). *World Dynamics*. Cambridge, MA: Wirght-Allen Press.

Frahm, M. (2011). *Beschreibung von komplexen Projektstrukturen*. PM Aktuell.

Frahm, M. (2015). *Kybernetisches Bauprojektmanagement Gestaltung lebensfähiger Baustrukturen auf Grundlage des Viable System Models*. Stuttgart: Booksondemand.

Frahm, M. and Rahebi, H. (2021). *Management von Groß- und Megaprojekten im Bauwesen – Grundlagen für eine komplexitätsgerechte Umsetzung von Infrastrukturvorhaben*. Springer.

Frandson, A., Berghede, K., and Tommelein, I. (2013). *Takt Time Planning for Construction of Exterior Cladding*. Fortaleza, Brazil: International Group for Lean Construction.

FTI-Andersch Consulting (2021). *German Engineering 2025: Welche Herausforderungen deutsche Maschinenbauer jetzt meistern müssen*. Berlin: FTI-Andersch Consulting.

Furukawa-Caspary, M. (2016). *Lean auf gut Deutsch: Band 1 Einführung und Bestandsaufnahme*. Norderstedt, Germany: Books on Demand.

German Lean Construction Institute (2019). *Lean Construction – Lean und Methoden*. Karlsruhe, Germany: German Lean Construction Institute.

Gigerenzer, G. (2008). *Bauchentscheidungen*. Goldmann.

Gorecki, P. and Pautsch, P. (2013). *Praxisbuch Lean Management*. Munich: Carl Hanser Verlag.

Greiman, V. (2013). *Megaproject Management Lesson on Risk and Project Management from the Big Dig*, 1e. Hoboken, NJ: Wiley.

Haghsheno, S. (2004). *Analyse der Chancen und Risiken des GMP-Vertrags bei der Abwicklung von Bauprojekten*. Mensch und Buch Verlag.

Haghsheno, S., Binninger, M., Dlouhy, J., and Sterlike, S. (2016). History and theoretical foundations of takt planning and takt control. 24th Conference of the International Group for Lean Construction, pp. 53–62. https://iglcstorage.blob.core .windows.net/papers/attachment-6aa12588-08a1-4f6b-8f82-4f51a463df98.pdf (accessed 22 March 2022).

Heidemann, A. (2011). *Kooperative Projektabwicklung im Bauwesen unter der Berücksichtigung von Lean-Prinzipien – Entwicklung eines Lean-Projektabwicklungssystems*. Karlsruhe, Germany: Universitätsverlag Karlsruhe.

Heller, T. and Prasse, C. (2018). *Total Productive Management: ganzheitlich: Einführung in der Praxis*. Berlin: Springer Vieweg.

Herlyn, W. (2012). *PPS im Automobilbau*. Munich: Carl Hanser Fachbuchverlag.

Herrera, J. and Munoz, A. (2015). First run video studies: driving continous improvement. https://leanconstruction.org/uploads/wp/media/docs/2015-congress/presentations/TH18%20-%20Herrera.pdf (accessed 22 March 2022).

Herrera, C., Thomas, A., Belmokhtar, S., and Pannequin, R. (2011). A viable system model for product-driven systems. International Conference on Industrial Engineering and Systems Management (IESM). Metz.

Herrmann, C., Bergmann, L., Halubek, P., and Thiede, S. (2008). Lean production system design from the perspective of the viable system model. In: *Manufacturing Systems and Technologies for the New Frontier* (ed. M. Mitsuishi, K. Ueda and F. Kimura). London: Springer.

Hetzler, S. (2008). "Brain supporting environments" für Entscheide in komplexen Systemen. Dissertation, Universität St. Gallen: Hochschule für Wirtschafts-, Rechts- und Sozialwissenschaften (HSG).

Hetzler, S. (2010). *Real-Time Control für das Meistern von Komplexität*. Frankfurt am Main: Campus Verlag GmbH.

Hoffmann, W. (2017). *Zum Umgang mit Komplexität. (Indikatorbezogenes Modell zur Bewertung von Komplexität in Bauprojekten)*. TU Kaiserslautern.

Horsager, D. (2013). *Vertrauen: Die Währung von morgen*. Kulmbach: Books4success.

Hounshell, D. (1984). *From the American System to Mass Production 1800–1932*. Baltimore, MD: The Johns Hopkins University Press.

Hoverstadt, P. (2009). *The Fractal Organization – Creating Sustainable Organisations with the Viable System Model*. Wiley.

Hoverstadt, P. (2018). *Viable System Model Course*. St. Hegne, Germany: Malik Alumni Network.

Hoverstadt, P. (2019). *Advanced Course – Viable System Model*. SCiO DACH Camp, Aalen.

Industrie- und Handelskammer für Niederbayern (2021). *Konjunkturbericht Jahresbeginn 2021*. Passau, Germany: Industrie- und Handelskammer für Niederbayern.

IPMA (2016). Complexity_sheet_de_v2.1 (IPMATabelle zur Bestimmung der Komplexität). https://www.p-m-a.at/pma-download/cat_view/38zertifizierung-ipma-level-a-d.html: IPMA (accessed 22 March 2022).

Jackson, C.M. (2003). *System Thinking, Creative Holism for Managers*. Chichester: Wiley.

Jacobson, H. and Roucek, J. (ed.) (1959). *Automation and Society*. New York: Philosophical Library, Inc.

JELBA Werkzeug & Maschinenbau GmbH & Co. KG (2015). *Unternehmensleitbild*. Hauzenberg-Jahrdorf, Germany: JELBA Werkzeug & Maschinenbau GmbH & Co. KG.

JELBA Werkzeug & Maschinenbau GmbH & Co. KG (2019/2020). *Datenbasis für Jahresabschluss*. JELBA Werkzeug & Maschinenbau GmbH & Co. KG.

Kamiske, G. and Brauer, J. (2011). *Qualitätsmanagement von A bis Z*. Munich: Hanser Verlag.

Kaplan, R. and Norton, D. (2003). *Strategy Maps: Converting Intangible Assets into Tangible Outcomes*. Cambridge, MA: Harvard Business School Publishing.

Keese, C. (2016). *The Silicon Valley Challenge: A Wake-Up Call for Europe*. Penguin Verlag.

Kippels, D. (1999). Bei Porsche regiert schlanke Produktion. https://www.ingenieur.de/technik/fachbereiche/produktion/bei-porsche-regiert-schlanke-produktion (accessed 22 March 2022).

Klepzig, H.-J. (2018). *Lean Management in der Praxis – Kritische Darstellung der Kernelemente und Erfolgsmessung*. Düsseldorf: Hans-Böckler-Stiftung.

Kostka, C. and Kostka, S. (2013). *Der Kontinuierliche Verbesserungsprozess*. Munich: Carl Hanser Verlag.

Kunze, S. (2019). Warum wir menschenähnliche Roboter ablehnen. https://www
.elektrotechnik.vogel.de/warum-wir-menschenaehnliche-roboter-ablehnen-a-
851070 (accessed 22 March 2022).

Küppers, E. (2019). *Eine transdisziplinäre Einführung in die Welt der
Kybernetik – Grundlagen, Modelle, Theorien und Praxisbeispiele*. Springer.

Kurzweil, R. (2021). *How to Create a Mind: The Secret of Human Thought Revealed*.
Viking Books.

Lassl, W. (2019/2020). *The Viability of Organizations*, vol. 1–3. Cham, Switzerland:
Springer Nature.

Lawrence, K. (2013). *Developing Leaders in a VUCA Environment*. Chapel Hill, NC:
UNC Executive Development – Kenan Flagler Business School.

Lean Construction Institute (2015). *Transforming Design and Construction – A
Framework for Change*. Arlington, VA: Lean Construction Institute.

Lean Enterprise Institute (2007). Breakthrough. https://www.slideshare.net/
cmarchwi/lean-breakthrough-timeline (accessed 22 March 2022).

Lechner, H. (2015). Vorschlag zur Einführung von Projektklassen.
ttp://www.hanslechner.at/index.php/download/69-projektklassen (accessed 22
March 2022).

Liker, J. and Convis, G. (2012). *The Toyota Way to Lean Leadership*. New York:
McGraw-Hill.

Liker, J. and Meier, D. (2006). *The Toyota Way Fieldbook: A Practical Guide for
Implementing Toyota's 4Ps*. New York: McGraw-Hill.

Locatelli, G., Mancini, E., and Romano, E. (2014). Systems engineering to improve the
governance in complex project environments. *International Journal of Project
Management* 32 (8): 1395–1410.

Luhmann, N. (1994). *Soziale Systeme*. Suhrkamp.

MacCormack, A., Rusnak, J., and Carliss, Y. (März, 2008). Exploring the duality
between product and organizational architectures: a test of the mirroring
hypothesis. https://www.hbs.edu/ris/Publication%20Files/08-039_1861e507-1dc1-
4602-85b8-90d71559d85b.pdf (accessed 22 March 2022).

Mack, O., Khare, A., Krämer, A., and Burgartz, T. (ed.) (2016). *Managing in a VUCA
World*. New York: Springer International Publishing.

Mahlamäki, K., Ström, M., Eisto, T., and Holtta, V. (2009). Lean product development
point of view to current challenges of engineering change management in
traditional manufacturing industries. In: *2009 IEEE International Technology
Management Conference (ICE)*. IEEE http://doi.org/10.1109/ITMC.2009.7461408.

Marquardt, C. (2018). Lean leadership. In: *Lean Construction – Das
Managementhandbuch* (ed. M. Fiedler), 465–485. Munich: Springer-Gabler.

Maturana, H. and Varela, F. (1992). *Tree of Knowledge: The Biological Roots of Human
Understanding*. Boulder, CO: Shambhala.

Meadows, D. (2017). *Thinking in Systems: A Primer*. Chelsea: Green Publishing.

Meadows, D., Randers, J., and Behrens, W. (1972). *The Limits to Growth; A Report for the Club of Rome's Project on the Predicament of Mankind*. New York: Universe Books.

Merrow, E. (2012). *Industrial Megaprojects: Concepts, Strategies and Practices for Success*, 1e. Hoboken, NJ: Wiley.

Merton, R.K. (1995). *Soziologische Theorie und soziale Struktur*. Berlin: de Gruyter.

Michalicki, M. and Schneider, M. (2020). *Kostenrechnung in der Lean Produktion: Verschwendung ausweisen, Wertschöpfung ermitteln, Entscheidungen verbessern*. Munich: Carl Hanser Verlag.

Miller, R. and Lessard, D. (2001). Understanding and managing risks in large engineering projects. *International Journal of Project Management* 19 (8): 437–443.

Miller, R. and Lessard, D. (2007). Evolving strategy: risk management and the shaping of large engineering projects. MIT Sloan School of Management MIT Sloan Working Paper (4639–07). https://dspace.mit.edu/handle/1721.1/37157 (accessed 22 March 2022).

Modig, N. and Ahlström, P. (2015). *Das ist Lean: die Lösung des Effizienzparadoxes*. Stockholm: Rheologica Publishing.

Moore, B., Calvo-Amodio, J., and Junker, J. (2015). Synthesizing systemic intervention approaches: combining viable system model, knowledge management, and Toyota production system for a sustainable holistic management model. 59th Annual Meeting of the International Society for the System Sciences (ISSS), pp. 1061–1079. Berlin: International Society for the System Sciences (ISSS).

Mossman, A. (2019). Using plus/delta for feedback and improving social processes. https://leanconstructionblog.com/Using-Plus-Delta-for-Feedback-and-Improving-Social-Processes.html (accessed 22 March 2022).

Mossmann, A. (2016). *Last Planner – 5+1 wichtige und kooperative Gespräche für eine zuverlässige Planungs- und Bausausführung* (trans. C. Nesensohn). Nottingham: Centre for Lean Projects, Nottingham Trent University.

Ohno, T. and Bodek, N. (1988). *Toyota Production System: Beyond Large-Scale Production*, vol. 2. Boca Rato, FL: Productivity Press.

Ohno, T. (2013). *Das Toyota Produktionssystem: Das Standardwerk zur Lean Production*. Frankfurt: Campus Verlag.

Parkes, A. (2015). Lean management genesis. *Management* 19 (2): 106–121.

Patzak, G.R. (2009). *Projektmanagement (Leitfaden zum Management von Projekten, Projektportfolios, Programmen und projektorientierten Unternehmen)*. Vienna: Linde.

Paul, H. (2020) Impuls Stiftung.de: Maschinen- und Anlagenbau schöpft neuen Mut. http://www.impuls-stiftung.de/viewer/-/v2article/render/57181145 (accessed 22 March 2022).

Pfeifer, T. and Schmitt, R. (2014). *Masing Handbuch Qualitätsmanagement*. Munich: Carl Hanser Verlag.

Pfeiffer, W. (1994). *Lean-Management: Grundlagen der Führung und Organisation lernender Unternehmen*. Berlin: Erich Schmidt Verlag.

Pfiffner, M. (2020). *Die dritte Dimension des Organisierens – Steuerung und Kommunikation*. Wiesbaden, Germany: Springer Fachmedien.

Pretting, G. (2006). Die Erfindung des Schlachtplans. https://www.brandeins.de/magazine/brand-eins-wirtschaftsmagazin/2006/kapitalismus/die-erfindung-des-schlachtplans (accessed 22 March 2022).

PWC, Niederdrenk, R., and Seemann, R. (2018). *Baubranche aktuell: Wachstum 2020: Digitalisierung und BIM*. Munich: PWC.

REFA AG (2021 A3-Methode/A3-Report. https://refa.de/service/refa-lexikon/a3-methode-a3-report (accessed 22 March 2022).

REFA AG (2022) Visuelles management. https://refa.de/service/refa-lexikon/visuelles-management (accessed 22 March 2022).

Reiß, M. (1982). Das Kongruenzprinzip der Organisation. *WiSt Wirtschaftwissenschaftliches Studium* 2: 75–78.

Reitz, A. (2009). *Lean TPM: In 12 Schritten zum schlanken Managementsystem*. Munich: mi-Fachverlag, FinanzBuch Verlag.

Roll, C. (2018). Organisation zur Selbstorganisation: Gestaltung eines wirksamen Lean-Management-Systems im mittelständischen Anlagen- und Maschinenbau mit Hilfe des Viable System Models. Master thesis, Donau-Universität Krems: Department für Wissens- und Kommunikationsmanagement; Zentrum für Kognition, Information und Management.

Rosenthal, M. (2002). The essence of jidoka. http://theleanthinker.com/wp-content/uploads/sites/3/2009/04/The-Essence-of-Jidoka-SME-Version.pdf (accessed 22 Marsh 2022).

Rother, M. (2013). *Die Kata des Weltmarktführers*. Frankfurt: Campus Verlag.

Rother, M. and Shook, J. (2011). *Sehen lernen – mit Wertstromdesign die Wertschöpfung erhöhen und Verschwendung beseitigen*. Mühlhein an der Ruhr, Germany: Lean Management Institut.

Schön, D. (2008). *The Reflective Practitioner: How Professionals Think in Action*. Basic Books.

Schwaninger, M. and Scheef, C. (2016). A test of the viable system model: theoretical claim vs. empirical evidence. *Cybernetics and Systems: An International Journal* 47 (7): 544–569.

Schwanninger, M. (2006). *Intelligent Organisations: Powerful Models for Systemic Management*. Springer.

Schwerdtner, P. (2007). *Anreizbasiertes Steuerungs- und Vergütungssystem für Einzelvergaben im Hochbau*. Institut für Bauwirtschaft und Baubetrieb der TU Braunschweig.

Senge, P. (2010). *The Fifth Discipline: The Art and Practice of the Learning Organization*. Cornerstone Digital.

Sharma, J. (2012). *Scrum und das Standardmodell wirksamen Managements nach Malik: Eine Synthese systemischen Managements.* Stuttgart: ibidem-Verlag.

Shingo, S. (1992). *Das Erfolgsgeheimnis der Toyota Produktion.* Landsberg/Lech, Germany: Verlag moderne Industrie.

Snowden, D. and Boone, M. (2007). A leader's framework for decision making. https://hbr.org/2007/11/a-leaders-framework-for-decision-making (accessed 23 March 2022).

Steiner, H. and Jernej, R. (2017). *Kybernetisches Bauprojektmanagement: Einige Grundlagen.* Steiner und Partner Schriften.

Steinhäusser, T., Fatos, E., Tommelein, I., and Lindemann, U. (2015). Management cybernetics as theoretical basis for lean construction. *Lean Construction Journal* 1–14.

Syska, A. (2006). Autonomation. In: *Produktionsmanagement: das A–Z wichtiger Methoden und Konzepte für die Produktion von heute*, 27–28. Wiesbaden, Germany: Gabler Verlag.

Takeda, H. (1995). *Das Synchrone Produktionssystem.* Landsberg/Lech, Germany: Verlag moderne Industrie.

Taleb, N. (2010). *The Black Swan: The Impact of the Highly Improbable.* New York: Random House Trade Paperbacks.

Toyota Material Handling (2021). Das Toyota Produktionssystem und seine Bedeutung für das Geschäft. https://www.tqu-group.com/we-dokumente/ Downloads/ToyotaPS.pdf (accessed 22 March 2022).

Toyota Motor Corporation (1998). *The Toyota Production System: Leaner Manufacturing for a Greener Planet.* Tokyo: TMC, Public Affairs Divison.

Toyota Motor Corporation (2021). Toyota Production System. https://global.toyota/ en/company/vision-and-philosophy/production-system (accessed 22 March 2022).

Ulrich, H. (2001). *Gesammelte Schriften: Werkausgabe in fünf Bänden.* Haupt.

Ulrich, H. and Probst, G. (1988). *Anleitung zum ganzheitlichen Denken und Handeln.* Haupt.

Verein Deutscher Ingenieure (2021). Ressourceneffizienz im Bauwesen. https://www .ressource-deutschland.de/themen/bauwesen (accessed 22 March 2022).

Verein Deutscher Ingenieure (2016). *Ressourceneffizienz – Methodische Grundlagen, Prinzipien und Strategien (VDI 4800 Blatt 1).* Düsseldorf: VDI-Gesellschaft Energie und Umwelt (GEU) – Fachbereich Ressourcenmanagement.

Verein Deutscher Ingenieure (2019). *Lean Construction (VDI 2553).* Düsseldorf: VDI-Gesellschaft Bauen und Gebäudetechnik (GBG) – Fachbereich Bautechnik.

VDMA Forum Industrie 4.0 & PTW Institut für Produktionsmanagement, Technologie und Werkzeugmaschinen (2018). *Leitfaden Industrie 4.0 trifft Lean: Wertschöpfung ganzheitlich steigern.* Frankfurt/Main: VDMA Verlag GmbH.

Velitchkov, I. (2020). *Essential Balances: Stop Looking and Start Seeing What Makes Organizations Work.* Kvistgaard.

Venkatraman, V. (2017). *The Digital Matrix: New Rules for Business Transformation Through Technology*. Boston: LifeTree Media.

Verlag moderne Industrie GmbH (2017). Heijunka: Gegen Marktschwankungen gewappnet. https://www.produktion.de/technik/heijunka-gegen-marktschwankungen-gewappnet-101.html (accessed 22 March 2022).

Voigt, K.-I. (2010). Lean management. https://wirtschaftslexikon.gabler.de/definition/lean-management-37747/version-140745 (accessed 22 March 2022).

Weis, C. (2013). Definition: lean management. http://www.business-on.de/lean-management-definition-lean-management-_id40863.html (accessed 22 March 2022).

Wiener, N. (2013). *Cybernetics: Or Control and Communication in the Animal and the Machine*. Eastford, CT: Martino Fine Books.

Winch, G. (2009). *Managing Construction Projects: An Information Processing Approach*. Chichester: Wiley.

Wirtschaftslexikon24 (2018). Taylorismus. http://www.wirtschaftslexikon24.com/d/taylorismus/taylorismus.html (accessed 22 March 2022).

Womack, J. and Jones, D. (1996). *Lean Thinking*. New York: Simon & Schuster.

Womack, J., Jones, D., and Roos, D. (1990). *The Machine that Changed the World: The Story of Lean Production*. New York: HarperCollins.

Womack, J., Jones, D., and Roos, D. (1992). *Die zweite Revolution der Automobilindustrie*. New York: Campus Verlag.

Zollondz, H.-P. (ed.) (2001). *Lexikon Qualitätsmanagement: Handbuch des modenen Managements auf der Basis des Qualitätsmanagements*. Munich: Oldenbourg Wissenschaftsverlag.

Glossary

Algedonic channel

Alarm signals that directly transmit positive or negative messages into the highest system (system 5). In the process, the usual reporting structure is bypassed unfiltered and crosses all management levels as a solitary signal. Whistle-blowers, for example, are a symptom of a nonfunctioning algedonic channel.

Archetypes/patterns

Generic structures of frequently observable behaviour patterns in systems. They are suitable for diagnosing, communicating, and solving problems.

Ashby's law

The law which says: 'Only variety can absorb variety.' This means that a complex environment needs an organisation capable of managing the complexity. But Ashby was often misunderstood. It should not be the case that management should have so much self-variety that it is no longer practicable. It is also important to choose a manageable environment.

Autopoiesis

Self-referential systems that refer to themselves, creating and maintaining themselves out of themselves. An example of an autopoietic system approach is the viable system model (VSM).

Beers' Triple

Beers developed dynamic performance measurements to measure the strategic dimension of an organisation and its operational dimension.

Building information modelling (BIM)

A digital method based on three- to n-dimensional object-oriented building models. The digital model of the building serves as an information source and data

Designing Intelligent Construction Projects, First Edition. Michael Frahm and Carola Roll.
© 2022 John Wiley & Sons Ltd. Published 2022 by John Wiley & Sons Ltd.

hub for the collaboration of project participants. At the centre of this is digital registration and the interconnection of all relevant data for illustrating the physical, functional, cost-, and time-related characteristics of a building.

Complexity

As a simple rule of thumb: complexity means that a system has many elements, relationships, and states which change over time – sometimes very rapidly. Complexity is subjective and depends on the context and on the observer. According to W. Ross Ashby, a measure of complexity is variety.

Cybernetics

According to Norbert Wiener, cybernetics concerns control and communication in the animal and the machine. Cybernetics is thus the overarching science that describes the control and regulation of machines, living organisms, and social systems.

Cynefin framework

A method to describe systems. The framework differentiates between simple, complicated, complex, chaotic, and disorientated systems.

Digital transformation

The change of whole industries due to digitalisation. Most information is available in digital form and can be used more effectively in the digital age and often in real time. Digital transformation can be separated into four categories: digital data collection and analysis, automation, networking, and digital access.

Digital twin

A virtual representation of a process, product, or service. The physical and virtual objects can be connected in almost real time with the help of sensors and data integration.

First-order principles

Unconditional truths or laws of nature that others accept and acknowledge. They have a higher degree of abstraction and are universally valid.

Flow efficiency

Indicates how efficiently a product flows through the organisation based on the customer's needs. It measures how long it takes to identify the customer's needs in order to meet them.

Fractal structure

Describes systems with similar elements (self-similarity) and is an essential rule in management cybernetics. The same or similar generic organisational code can deal with complexity in a relatively simple way and help create order. Fractal structures are also found in mathematics (e.g. Fibonacci and Mandelbrot-set), art, and nature.

Intelligence

Adaptability and robustness in organisations and low-waste processes that promote the application of new technologies in a sustainable way with an eye on the entire production system and its alignment and coupling internally and externally. Intelligence also stands for cooperative collaboration and production systems in which it is fun to work, and in which there is a working culture of motivation and a responsible and sustainable use of resources, as well as the use of technical and digital systems to relieve the human workforce of tasks that can be done more sensibly, purposefully, and economically.

Last planner system (LPS)

Aims to stabilise planning and control processes during production. Obstacles are to be overcome with the help of foresighted and short-cycle planning. It is a holistic and collaborative approach to increase productivity.

Lean management

Derived from the Toyota Production System and focused on adding value for the customer. Added value is defined by what the customer is willing to pay. Hence, lean management focuses on the elimination of waste of all kinds.

Management cybernetics

Applies the principles of cybernetics to organisations and the business world. Developed by Stafford Beer, management cybernetic approaches try to increase the viability of organisations. One application of management cybernetics is the viable system model (VSM).

Nudge management

Stimulates the instincts of individuals. The nudges form the framework conditions and ensure employees act according to the desired culture without thinking about it. What is important here is that employees have freedom of choice and do not feel obliged to do something.

Purpose

According to Stafford Beer, 'The purpose of the system is what it does.' This corresponds to the purpose/business vision and mission of an organisation. Accordingly, large organisations can have several purposes.

Recursion levels

Organisational units of systems (here: companies or project organisations) that are either integrated into each other or built on each other. An essential characteristic of these recursion levels is a certain self-similarity of the individual structural elements. A recursion level is a management level in an organisation that includes all levels below it and can also describe where an organisational unit is structurally located in the overall organisation.

Resource efficiency

Means that an organisation uses the required resources (facilities, personnel, tools, information systems) during the production process to achieve the maximum utilisation of resources.

Second-order principles

Second-order principles formulate instructions for execution and contain statements such as: 'You should… to achieve this'. In contrast to first-order principles, they have a higher practical relevance.

Standardisation

Elementary for detailed production planning and crucial for quality assurance. Standards reduce the variety in the production process and therefore decrease complexity. Standardised workflows and parts ensure a high level of quality and stable processes and allow for high scalability.

System

Consists of elements, their relationships and their behaviour, and its purpose and function. It interacts with itself and its environment. Through interrelationships, the whole is something other or more than just the sum of its parts, for example a single human being, a soccer team, a company, a city, an economy, an animal, a tree, a forest, the earth, the solar system, as well as the galaxy. Conglomerates without certain connections or functions are not systems.

System 1 – Operation/system in focus

System 1 consists of three elements: (i) environment, (ii) operation, and (iii) management. It is the reason why the organisation exists.

System 2 – Coordination

System 2 has two tasks. First, it represents a communication medium between S1 and S3 through standardised processes and coordinates all S1 systems. Second, it is the institutional place where self-organisation takes place. It has an activating or 'sympathetic' effect.

System 3 – Operational management

System 3 deals with systems 1 to 3 and must make all operational activities as efficient as possible. It allocates resources and demands results. S3 competes with S4 for resources and receives normative specifications from system 5.

System 3* – Monitoring/audit

System 3* is a review channel with an institutional absorbing or 'parasympathetic' effect. It provides system 3 with additional information for an overall picture of the organisation or for a broader perspective.

System 4 – Strategic management

System 4 focuses on the strategic issues of the overall organisation and deals with future problems from the environment. It competes with S3 for resources.

System 5 – Policy

System 5 represents the identity of the organisation. Topics such as values, norms, ethics, and culture are addressed here and transferred into the organisation. The highest decision-making unit makes fundamental choices and, if necessary, regulates between S3 and S4.

System dynamics

Serves the holistic analysis and simulation of complex and dynamic systems. System dynamics is a subscience of systems theory/thinking. It can identify dynamic, complex, and nonlinear problems and model solutions. Through modelling, the behaviour of the systems can be analysed and understood, which ultimately leads to well-founded decisions.

System thinking

Also known as networked thinking. It means thinking in cycles of action, networks, and interrelations and applying system thinking methods, knowledge of archetypes, and thinking in models.

Takt

The takt forms the heartbeat of production and indicates the speed at which work steps are carried out. The application is particularly suitable for frequently manufacturing recurring or similar operations or sequence sections (e.g. carriageways, tunnels, high-rise buildings, bridges).

Takt control

During the construction process, the construction site is coordinated through takt control and includes takt control meetings and a takt board. Takt control meetings are held daily to reflect the previous day's performance and to discuss upcoming work packages. The control board makes it possible to share important information with those involved in the project.

Takt planning

It is a method to standardise and create a consistent workflow in production. Adapted on construction projects, takt planning includes the following steps:

1. Zoning of the construction project.
2. Definition of workload.
3. Development of a trade sequence.
4. Definition of takt time and takt area.
5. Adjusting activities.

6. Determining takt strategy and production schedule.
7. Repetition of the steps and creation of a takt plan.

Technical singularity

The chief engineer of Google, Ray Kurzweil, predicted that technical singularity, which will completely restructure human life, will be reached by 2045. Technical singularity means that, owing to the exponential progress of innovation and the resulting technologies, the future of technological development is no longer predictable.

Tit for tat

Tit for tat ('Like you to me, like I to you') is an approach from game theory. The winner of the long-term game is always the one who starts with a cooperative action and then adapts to the behaviour of their opponent. The success of this cooperation strategy was proven in an experiment by Robert Axelrod.

Transducers

Converters that form the interface between subsystems. They ensure the maintenance of information authenticity.

Types of waste

The seven/eight typical types of waste in lean management are:

- Transportation
- Inventory
- Movement
- Waiting
- Overproduction
- Overprocessing
- Defects and rework
- Skills.

Mnemonic: **TIM WOODS** – initial letters of the eight typical types of waste.

Variety

According to W. Ross Ashby, variety is the central measure of complexity in cybernetics or management cybernetics. It describes the number of states a system can assume. Variety depends on the number of elements, their relationships, their behaviour, and their change over time in a system (see also Ashby's law).

Viable system model (VSM)

Used in management cybernetics and developed by Stafford Beer. Beer oriented its development on the successful model of evolution in terms of viability: the central nervous system of mammals. The model uses six subsystems and includes consideration of the environment.

List of Figures

Designing Intelligent Construction Projects, First Edition. Michael Frahm and Carola Roll.
© 2022 John Wiley & Sons Ltd. Published 2022 by John Wiley & Sons Ltd.

List of Tables

Designing Intelligent Construction Projects, First Edition. Michael Frahm and Carola Roll.
© 2022 John Wiley & Sons Ltd. Published 2022 by John Wiley & Sons Ltd.

List of Equations

Designing Intelligent Construction Projects, First Edition. Michael Frahm and Carola Roll.
© 2022 John Wiley & Sons Ltd. Published 2022 by John Wiley & Sons Ltd.

List of Abbreviations

2D	two-dimensional
3D	three-dimensional
5A/5S	sort – set in order – shine – standardise – sustain
5W/6W	5 why/6 why
A3	quality management tool for continuous improvement
AI	artificial intelligence
ANP	analytical network process
AR	augmented reality
ASME	American Society of Mechanical Engineers
BIM	building information modelling
BSC	balanced scorecard
CIP	continuous improvement process/Kaizen
CNC	computerised numerical control
CPS	cyber physical system
D	disturbance
DIN	German Institute for Standardisation
DSGVO	German general data protection regulation
EDP	electronic data processing
EN	European standard
ERP	enterprise resource planning
ESTW	electronic signal boxes
GLCI	German Lean Construction Institute
GMP	guaranteed maximum price
GPEX	business process excellence
GPS	global positioning system
HVAC	heating, ventilation, air conditions
IGLC	Conference of International Group for Lean Construction
IoT	Internet of things
IPD	integrated project delivery

Designing Intelligent Construction Projects, First Edition. Michael Frahm and Carola Roll.
© 2022 John Wiley & Sons Ltd. Published 2022 by John Wiley & Sons Ltd.

IPMA	International Project Management Association
ISO	International Organisation for Standardisation
JIS	just in sequence
JIT	just in time
KPI	key performance indicator
KTA	Nuclear Committee in Germany
L	letter
LCI	Lean Construction Institute
LP	last planner
LPA	layered process audit
LPS	last planner system
MIT	Massachusetts Institute of Technology
MR	mixed reality
MVP	minimal viable products
OEE	overall equipment effectiveness
PDA	production data acquisition
PDCA	plan, do, check, act
PFA	planning approval section
PO	project organisation
POSIWID	The purpose of the system is what it does
PPC	planned percent complete
PRM	people with reduced mobility
PS	project controlling team
PTW	Institute for Production Management, Technology and Machine Tools
QM	quality management
R	regulation
REFA	Association for Work Design, Business Organisation and Corporate Development
RFID	radio-frequency-identification
RWTH	Rhenish-Westphalian Technical University
S1	System 1
S2	System 2
S3	System 3
S3*	System 3*
S4	System 4
S5	System 5
SCC	safety certificate contractors
SCiO	system and complexity in organisation
SGMM	St. Gallen Management Model
SMED	single minute exchange of die

SPM	subproject manager
TIM WOODS	transportation, inventory, movement, waiting, overproduction, overprocessing, defects and rework, skills
TISAX	Trusted Information Security Assessment Exchange
TPS	Toyota Production System
USB	universal serial bus
VDI	Association of German Engineers
VDMA	German Mechanical and Plant Engineering Association
V_c	variety of consequences
V_d	variety disturbance
V_e	variety environment
V_m	variety management
VOB	German construction contract procedures
V_r	variety reaction
VR	virtual reality
VSM	viable system model
VUCA	volatility – uncertainty – complexity – ambiguity and vision – understanding – clarity – agility
WZL	Machine Tool Laboratory

Index

Designing Intelligent Construction Projects, First Edition. Michael Frahm and Carola Roll.
© 2022 John Wiley & Sons Ltd. Published 2022 by John Wiley & Sons Ltd.